Lecture Notes in Computer Science 13744

More information about this series at https://link.springer.com/bookseries/558

Chen Yu · Jiehan Zhou · Xianhua Song ·
Zeguang Lu (Eds.)

Green, Pervasive, and Cloud Computing

17th International Conference, GPC 2022
Chengdu, China, December 2–4, 2022
Proceedings

Springer

Editors
Chen Yu
Huazhong University of Science
and Technology
Wuhan, China

Xianhua Song
Harbin University of Science and Technology
Harbin, China

Jiehan Zhou
University of Oulu
Oulu, Finland

Zeguang Lu
National Academy of Guo Ding Institute
of Data Science
Beijing, China

ISSN 0302-9743   ISSN 1611-3349 (electronic)
Lecture Notes in Computer Science
ISBN 978-3-031-26117-6   ISBN 978-3-031-26118-3 (eBook)
https://doi.org/10.1007/978-3-031-26118-3

This Springer imprint is published by the registered company Springer Nature Switzerland AG
The registered company address is: Gewerbestrasse 11, 6330 Cham, Switzerland

Preface

We are pleased to welcome you to read the proceedings of the 17th International Conference on Green, Pervasive, and Cloud Computing (GPC 2022). It was held in Chengdu, China on December 2–4, 2022. GPC 2022 provided a forum for researchers and practitioners to discuss and exchange new ideas, work-in-process, research and practice in Green, Pervasive, and Cloud Computing.

The Program Committee received 104 submissions in total. After careful reviews, we accepted 19 papers for publication in the conference proceedings with an acceptance rate of 18.3%.

We would like to thank the members of the Program Committee from institutes across the world for their hard review work. Their collective efforts and diverse expertise made the conference program successful and exciting. Their review feedback and comments were really helpful for the authors to improve their papers and research.

We would like to thank all the authors and participants for their great support in making the conference successful.

We thank the team from Springer for their professional assistance in the publication of the conference proceedings.

December 2022
Chen Yu
Jiehan Zhou

Organization

General Chairs

Zhiguang Qin — University of Electronic Science and Technology of China, China

Rajkumar Buyya — The University of Melbourne, Australia

Program Chairs

Chen Yu — Huazhong University of Science and Technology, China

Jiehan Zhou — University of Oulu, Finland

Publicity Chair

Fangming Liu — Huazhong University of Science and Technology, China

Web Chair

Qianqian Wang — Huazhong University of Science and Technology, China

Publication Chairs

Xianhua Song — Harbin University of Science and Technology, China

Zeguang Lu — National Academy of Guo Ding Institute of Data Science, China

Steering Committee Chair

Hai Jin — Huazhong University of Science and Technology, China

Steering Committee

Nabil Abdennadher — University of Applied Sciences, Switzerland

Christophe Cerin — University of Paris XIII, France

Sajal K. Das	Missouri University of Science and Technology, USA
Jean-Luc Gaudiot	University of California-Irvine, USA
Kuan-Ching Li	Providence University, Taiwan
Cho-Li Wang	The University of Hong Kong, China
Chao-Tung Yang	Tunghai University, Taiwan
Laurence T. Yang	St. Francis Xavier University, Canada, and Hainan University, China
Zhiwen Yu	Northwestern Polytechnical University, China

Workshop Chairs

Ning Zhang	University of Windsor, Canada
Zhicai Zhang	Shanxi University, China
Zhe Zhang	Nanjing University of Posts and Telecommunications, China
Yongliang Qiao	The University of Sydney, Australia
Daobilige Su	China Agricultural University, China
Zichen Huang	Kyoto University, Japan
Yangyang Guo	Anhui University, China
Qiankun Fu	Jilin University, China
Honghua Jiang	Shandong Agricultural University, China
Meili Wang	Northwest A&F University, China
Fenghua Zhu	Institute of Automation, Chinese Academy of Sciences, China
Ting Xu	Chang'an University, China
Ryan Wen Liu	Wuhan University of Technology, China
Rummei Li	Beijing Jiaotong University, China
Tigang Jiang	University of Electronic Science and Technology of China, China
Shaoen Wu	Illinois State University, USA
Qing Yang	University of North Texas, USA
Kun Hua	Lawrence Technological University, USA
Guangjie Han	Hohai University, China
Kai Lin	Dalian University of Technology, China
Xinguo Yu	Central China Normal University, China
Jiehan Zhou	University of Oulu, Finland
Jun Shen	Southeast University, China
Wenbin Gan	National Institute of Information and Communications Technology, Japan
Keke Huang	Central South University, China

Contents

SFYOLO: A Lightweight and Effective Network Based on Space-Friendly Aggregation Perception for Pear Detection

Yipu Li[1,2,3], Yuan Rao[1,2,3(✉)], Xiu Jin[1,2,3], Zhaohui Jiang[1,2,3], Lu Liu[1,2,3], and Yuwei Wang[1,2,3]

[1] College of Information and Computer Science, Anhui Agricultural University, Hefei 230036, China
raoyuan@ahau.edu.cn
[2] Key Laboratory of Agricultural Sensors, Ministry of Agriculture and Rural Affairs, Hefei 230036, China
[3] Anhui Provincial Key Laboratory of Smart Agricultural Technology and Equipment, Hefei 230036, China

Abstract. It is always challenging for efficiently conducting accurate detection of small and occluded pears in modern orchards. In the past few years, the aforementioned detection tasks remained unsolved though lots of researchers attempted to optimize the adaption of background noise and viewpoints, particularly compliant models suitable for simultaneously detecting small and occluded pears with low computational cost and memory usage. In this paper, we proposed a lightweight and effective object detection network called as SFYOLO based on space-friendly aggregation perception. Specifically, a novel space-friendly attention mechanism was proposed for implementing the aggregate perception of spatial domain and channel domain. Afterwards, an improved space-friendly transformer encoder was put forward for enhancing the ability of information exchange between channels. Finally, the decoupled anchor-free detectors were used as the head to improve the adaptability of the network. The mean Average Precision (mAP) for in-field pears was 93.12% in SFYOLO, which was increased by 2.03% compared with original YOLOv5s. Additional experiments and comparison were carried out considering newly proposed YOLOv6 and YOLOv7 that aimed at optimizing the detection accuracy and speed. Results verified that small and occluded pears could be detected fast and accurately by the competitive SFYOLO network under various viewpoints for further orchard yield estimation and development of pear picking system.

Keywords: YOLOv5s · Object detection · Visual attention mechanism · Transformer encoder · Aggregate perception

1 Introduction

To meet the consumption requirements of the world's growing population, horticulture has been trying to find new ways to increase orchard productivity [1,2].

C. Yu et al. (Eds.): GPC 2022, LNCS 13744, pp. 1–16, 2023.
https://doi.org/10.1007/978-3-031-26118-3_1

The development of artificial intelligence and robot technology provides a feasible scheme for the improvement of production efficiency and efficiency [3,4]. However, in the practical application scenario of agriculture, there are still great challenges in the application of the above technologies. Although labor-intensive agronomic management based on manual labor is difficult to meet the needs of agricultural development under the background of rising agricultural costs and shortage of skilled labor, however, the technology-intensive agronomic management dominated by computer science and technology is still lack of practical experience. If the transition from labor-intensive orchard to technology-intensive orchard can be realized, the development of automatic agronomic management, such as pear growth monitoring, yield estimation and automatic picking of fruits, will help in reducing economic and environmental costs.

In the latest development of intelligent agriculture and independent production, deep learning technology has been widely used practically in agricultural management to improve and estimate agricultural production. Traditionally, it is common for machine vision methods based on convolutional neural networks (CNN) to implement fruit detection. However, due to various sizes from big to small, severe occlusion caused by dense distribution, different viewpoints captured by different device, and several other factors result in obstacles and restrictions of object detection with satisfactory accuracy. Taking the detection of small pear as the main task, how to accurately detect small fruits and severely occluded fruits in complex orchard environments is the key to realize fruit growth monitoring and intelligent yield estimation. For the purpose of improving detection accuracy, current detection networks generally tend to increase the depth of the neural network and use dense connections [5,6], but this may lead to feature loss for small fruits. Therefore, an appropriate detection method is still required for implementing agronomic management in an effective and efficient way.

In order to overcome the above shortcomings, based on the existing YOLO series networks [7–11], we strive to balance the detection accuracy and computational cost, and design a Space-Friendly YOLO (SFYOLO) network. Firstly, a novel SFA (Space-Friendly Attention) mechanism was designed, which enabled the aggregation of spatial and channel attention at a low computational cost. Subsequently, in order to improve the overall perception ability of the SFYOLO, we introduced transformer encoder as the global feature extraction and modeling module, and the SFA mechanism was embed into it to build SF-TE (Space-Friendly Transformer Encoder). Afterwards, SF-TE was used as the feature extraction unit of the neck part of the network to re-extract multi-level features. Finally, the decoupled anchor-free detectors were used as the head of the network to improve the adaptability and the accuracy of the detection network, especially for the dense regions of small and occluded pears.

2   Related Works

Self-Attention in Computer Vision. Visual attention mechanism is derived from studies of the human vision system. Different parts of the human retina have different levels for information processing. In order to make rational use of limited visual information processing resources, humans usually focus on specific parts of the visual area. Generally speaking, the attention mechanism determines which part of the input needs to be paid more attention to. The ability and efficiency of feature extraction will be improved if the network focuses only on task-related regions, rather than useless regions. Self-Attention mechanisms, such as SE [12], CBAM [13] and CA [14], assign different weights to the region that need attention and eliminate irrelevant information, thus improving the quality of the extracted features.

Vision Transformer. Due to the great success of Transformer in the field of NLP (Natural Language Processing), lots of research has been done trying to transfer it to the field of computer vision. Vision Transformer [15] firstly proves that the images can be directly applied in Transformer, it makes one image into sequence of patches and reaches SOTA performance in large datasets than traditional CNN-based networks. DETR [16] is the first network based on Transformer for object detection task. It simplifies the process of target detection by treating the detection problem as ensemble prediction, offering an effective way of combining CNN with Transformer. However, the limitations of two networks above are that they both require large-scaled datasets and take too much time for training. Some networks such as LeViT [17], CvT [18], and Visformer [19] alleviate these problems by means of multi-scale feature fusion. The aforementioned work has demonstrated that the proper combination of CNN and Transformer can help in reducing the inference time and the network size, enabling it to be applied in the real-time object detection task.

Object Detection Models. With the development of deep learning, various object detection networks and methods have been proposed. The existing object detection methods can be divided into two categories: 1) One-stage detectors, such as YOLO series, FCOS [20], SSD [21]. 2) Two-stage detectors, such as VFNet(17), Faster RCNN(18). In recent research, several novel one-stage object detection networks, e.g. YOLOv6 [22] and YOLOv7 [23], were proposed to further improve the object detection performance in general scenarios. However, the computational cost increased significantly due to the use of high-capacity feature extraction backbone networks and feature extraction modules, which made it difficult to be directly applied to real-time detection tasks in agricultural scenarios with limited resource of computing hardware. At the same time, the overall performance of YOLOv6 and YOLOv7 in detecting small and occluded pears remained to be verified.

3   Image Acquisition and Processing

3.1   Image Dataset Acquisition

The pear dataset was collected in September 2021 from the experimental pear orchard located in Suzhou City, Anhui Province, China. We divided the data collection work into two parts: collecting ground photography dataset and aerial photography dataset. The CCD camera mounted on the tripod and the CCD camera mounted on the Unmanned Aerial Vehicle (UAV) were used to capture pear images from the ground and the air, respectively. A total of 1840 pear images were collected at a shooting distance of about 2 m to form a ground dataset. The aerial data set was collected by UAV at the height of 2 m above the pear tree canopy, and 985 images were obtained to form the aerial dataset. The collected images had a dense distribution of target pears of different sizes and contain a large number of target pears with different distances and shapes, which brought a great challenge to accurate detection.

3.2   Annotation Procedure and Data Augmentation

We annotated the ground 1840 images and the annotation information was saved in COCO format [24]. During data annotation, the percentage of small targets was especially increased, and each image contained up to 100 annotation boxes. The images were resized from 1280×720 pixels to 640×384 pixels with some pears covered by only a few pixels, which further increased the detection difficulty. The training of deep learning usually required a large amount of data. As we know, the limited data collected in real scenarios was often insufficient for network training. Therefore, we expanded the training samples by data augmentation for improving the generalization ability and robustness of the network. In this paper, we used random left-right flip, random up-down flip, HSV space transformation, random blur, Mosaic enhancement [9], Mixup enhancement [25], and other image preprocessing approaches provided in YOLOv5s for performing online enhancement of the data during training to expand the training set.

4   SFYOLO Network Design

4.1   Overall Structure of SFYOLO

Although YOLOv5 has been widely used in various fields, there are still some problems remained to be solved. First of all, although YOLOv5s had a lighter network architecture and faster detection speed than majority of networks, YOLOv5s had a certain sacrifice in accuracy. Thus, how to make up for the lack of detection accuracy without significantly improving the detection speed is an urgent problem to be solved. Secondly, the probe head of YOLOv5s adopted no anchor frame structure, although it performed well in the general scene, but in the scene where a large number of small and occluded pears need to be detected, YOLOv5s was prone to suffering from missed and false detection. Developing appropriate methods that could improve the detection ability of these hard-to-difficult pears would definitely promote the transformation of the network from theory to practical application. In order to facilitate the deployment on the intelligent picking platform in the agricultural scene and reduce the computing power and storage space requirements of the network as much as possible, we followed the general architecture of YOLOv5 to meet the real-time requirements.

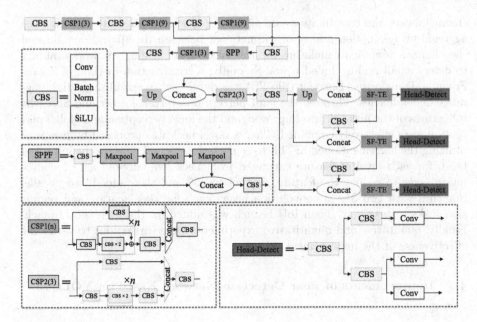

Fig. 1. The network architecture of SFYOLO (where SF-TE corresponds to space-friendly transformer encoder in Fig. 3)

The overall structure of SFYOLO is shown in the Fig. 1, and its framework can be summarized into three main parts, namely, the backbone, neck, and head. The backbone extraction network is CSPDarknet. After loading the pre-training weights on the COCO dataset, it can extract the necessary feature information from the original three-channel input image for subsequent detection and classification tasks. The neck PANet was used to reprocess the multi-scale feature images extracted by backbone at different stages. The basic feature extraction module was replaced with SF-TE, which could better perceive local and non-local aggregate features. The main part of head consists of three detectors. We chose anchor-free detectors instead of the original anchor-based detectors to improve the generalization ability of the network, with potential improvement in detection of small and occluded pears. At the same time, the detectors were decoupled and the classification process of frame and category was separated, which not only greatly improved the convergence speed, but also increased the classification and localization performance of the head.

4.2 Technical Route of Pear Detection Network

The flow of the SFYOLO-based pear detection network proposed in this study could be concluded as follow. Firstly, improvements were made in the neck and head of the original YOLOv5 network. For purpose of improving the neck, we proposed a Space-Friendly Attention (SFA) mechanism based on the channel-aware and coordinate-aware aggregation, which aimed at enhancing the aggregation perception ability of the network between the spatial domain and the channel domain to make up for the lack of the original network's ability to detect small and occluded pears. Secondly, a construction method of Space-Friendly Transformer Encoder (SF-TE) was developed to establish spatial long-range dependencies. Through the joint perception of Transformer and SFA, the robustness of the network was improved, and the local perception and global perception were effectively combined. Then a novel neck was proposed by means of utilizing the aforementioned SF-TE to replace the original CSP. In the improved head, for each level of feature extracted from neck, we adopted a 1×1 convolution layer to reduce the feature channel to 256 and then added two parallel branches with two 3 × 3 convolution layers each for classification and regression tasks respectively. Then IoU branch was added on the regression branch. Finally, qualitative and quantitative experiments were carried out to verify the effectiveness of the improvements.

4.3 Improvements of Pear Detection Network Based on YOLOv5s

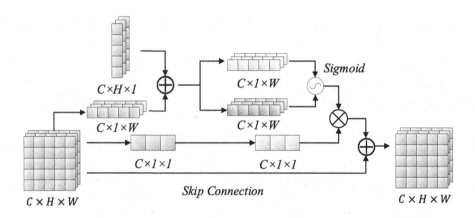

Fig. 2. The structure of space-Friendly Attention mechanism (SFA)

Space-friendly Attention Mechanism (SFA). In machine vision tasks, spatial information was equally important as coordinate information. Achieving an aggregate perception of spatial and channel features was beneficial to the overall perception ability of the network. Self-attention mechanism within both spatial and channel domains of the feature maps contributed to capturing dimensionally richer information in local and global way. As illustrated in Fig. 2, on the basis that the coordinate attention mechanism summarized the feature map into a pair of feature vectors along the horizontal coordinate direction and vertical coordinate direction, we added spatial feature vectors along the channel direction to implement the exchange of information between channels. This set of spatial feature vectors could selectively enlarge the valuable feature channels and restrain the useless feature channels, thus improving the performance of the network. Each feature map was aggregated into 1×1 pixels, which were reweighted with horizontal and vertical feature vectors to achieve an effective aggregate perception of spatial and channel features. In order to reduce the loss of original information and alleviate the problem of gradient disappearance, we artificially made some layers of the network skip the connection of the next layer of neurons, making the non-adjacent layers connected, weakening the strong connection between each layer, and alleviating the degradation of network caused by the excessive depth. By making full use of spatial and channel information, it was expected to improve the ability to capture local and global information in the network.

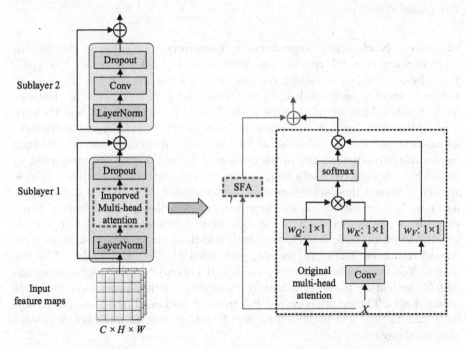

Fig. 3. The structure of space-friendly transformer encoder (SF-TE), where the red marks the improvements, SFA corresponds to Fig. 2 (Color figure online)

Space-friendly Transformer Encoder (SF-TE). Inspired by the idea of Visformer to introduce convolution into transformer encoders, we designed space-friendly transformer encoders. The structure of improved space-friendly transformer encoder is illustrated in Fig. 4, which could be divided into two sublayers, with the first layer being a multi-head attention layer and the second sublayer being a fully connected layer. Specifically, the input feature maps first passed through a multi-head attention sublayer, in which multiple attention heads performed attention computations synchronously to establish long-range dependencies in the spatial domain and preserve non-local features. Then they passed through the fully connected layer, where the feature maps were re-stored to their original size by multiple convolution operations. Finally, the result of residual connection was output after the Dropout operation, contributing to better convergence and overfitting prevention. To further achieve the proper perception of high-density targets, occluded targets, and small targets, the space-friendly attention mechanism was embedded into the multi-head attention mechanism. On the basis of the original multi-head attention mechanism, we added a path for feature extraction through SFA and connected it with the feature map extracted by the original multi-head attention mechanism in a cross-layer way. This expanded the channel dimension of the network, thus making full use of the channel information and achieving the aggregate perception in both channel and spatial domains.

Improved Neck with Transformer Encoders. To enhance the feature extraction ability of the network for target objects of different scales, YOLOv5s generally consisted of five stages referred to as [P1, P2, P3, P4, P5]. The output feature maps of these stages had distinct scales and were used to extract features from objects of different sizes, finally, the feature maps of P3, P4, and P5 were used for the detection task. These feature maps at different scales provided extensive multi-scale feature information for the target detection task. To enhance the feature extraction ability of the network for small targets, we attempted to embed SF-TE (space-friendly transformer encoder) into the neck. Since SF-TE greatly increased the parameters and computation cost of the network, applying it on low-resolution feature maps instead of high-resolution feature maps would lead to the increment of the expensive computation and memory cost, which was an obstacle for implementing real-time target detection tasks with limited resources. Therefore, we only embedded SF-TE into P3, P4, P5 in the neck of YOLOv5s. The improved Neck effectively enriched the information content of the feature maps, significantly enhancing the over perception results of in-field pears. The eventual output feature map contained denser non-local and local information, which allowed for the detection of small and occluded pears in natural environments.

Decoupled Anchor-Free Detectors. The detector of object detection network can be divided into anchor-based and anchor-free, and the former is usually used in traditional target detection network. However, the detection performance of anchor-based depends on the design of anchor frame to some extent, and it is very sensitive to the size, aspect ratio and quantity of anchor box. However, a large number of hyperparameters are used in the initialization design of the anchors, which makes it difficult to adjust these hyperparameters and cost a lot of time and computation to optimize. Considering that there were a large number of small pears in this application scene, which was different from the size distribution of the target in the general scene, it could not well match the preset anchor size. Therefore, we chosen to use the non-anchor detector instead of the anchor-based detector. Some scholars argued that there is a spatial misalignment problem in localization and classification tasks of object detection [26]. This meant that the two tasks have different focuses and degrees of interest, with classification focusing more on the features extracted that are most similar to existing categories, while localization focuses more on the location coordinates with the anchor points and thus on the bounding box parameter correction. Therefore, we chosen to use a decoupling head structure that can independently complete classification and positioning tasks by using different decoupled head branches, which is beneficial to the final detection effect and accuracy.

5 Experiments

5.1 Evaluation Metrics

Since the images collected in the natural environment contained many pears of different sizes, we used AP and F1-score to evaluate the performance of the network. AP refers to the average value of all ten intersections (IoU) thresholds with a uniform step size of 0.05 in the range of [0.50, 0.95]. F1-score is the harmonic average of precision and recall, with a value ranging from 0 to 1. Among them, a large F1-score value indicates good detection accuracy. The formula for the F1-score is as follows:

$$P = TP/(TP + FP) \tag{1}$$

$$R = TP/(TP + FN) \tag{2}$$

$$F = \frac{2 \times P \times R}{(P + R)} \tag{3}$$

where TP (True Positive) denotes the number of predicted positive samples, FP (False Positive) denotes the number of predicted positive but negative samples, and FN (False Negative) denotes the number of predicted negative but positive samples.

5.2 Implementation Details

The network was trained on the ground dataset with eight images as a batch and the loss was updated once per iteration for a total of 200 epochs on a single NVIDIA GTX 2080Ti. The detection of pears was carried out on a single NVIDIA GTX 1650, which simulated the limited computing environment in the natural environment. Using SGD as the optimizer, the initial learning rate was set to 0.01, the weight decay rate was set to 0.00048, and the momentum factor was set to 0.937. It gradually decayed to 1E-4 as the iterations proceeded. This experimental network was trained using transfer learning, and further training was performed on the pre-trained weights of the MS COCO dataset.

5.3 Quantitative Experimental Results Compared with YOLO Series

In addition to the proposed SFYOLO, the typical or recent members of YOLO series, such as YOLOv3-SPP, YOLOv4s, YOLOv5s, YOLOv6s and YOLOv7s were employed for making performance comparison based on the ground dataset. The detection accuracy, detection time, memory usage was taken into consideration. From Table 1, it could be seen that the proposed SFYOLO achieve on AP (average precision) and F1-score of 93.12% and 86.73%, which was much better than the results acquired by YOLOv3-SPP, YOLOv4s and YOLOv5s. Contrastively, YOLOv6s and YOLOv7s could offer closer detection results to the proposed SFYOLO. However, compared to the aforementioned two networks, the proposed SFYOLO reduced the detection time by 27% and 40%, meanwhile, the FLOPs decreased by 58% and 82.6%. In addition, in comparison to YOLOv5s, the AP of SFYOLO increased by 2.03% with slightly higher detection time and a larger memory usage. Therefore, it could be drawn that the proposed SFYOLO had the ability of offering better results in pear detection.

Table 1. Comparison among SFYOLO and other widely used and novel networks of YOLO series in terms of detection accuracy and efficiency on the ground dataset.

Networks	AP(%)	F1-score(%)	Detection time (ms)	Memory usage(MB)	FLOPs(G)
YOLOv3-SPP	89.51	83.72	16.1	120.32	157.1
YOLOv4s	91.98	85.62	20.2	246.34	137.2
YOLOv5s	91.09	84.35	**11.2**	**13.70**	**16.4**
YOLOv6s	92.93	85.92	18.2	285.78	44.2
YOLOv7s	93.02	86.23	22.3	64.23	104.7
SFYOLO	**93.12**	**86.73**	13.2	54.52	18.2

5.4 Qualitative Experiments Results on Ground Dataset

When detecting pears in practical application scenes, the collected images often contain a large number of pears with different scales, serious occlusion and cluttered density. It was difficult to detect these pears, which was often the main factor leading to low detection accuracy. If the improvements of the proposed network could effectively enhance the detection accuracy of these pears, it could provide a feasible scheme for the application of deep neural network in the field of agriculture. In order to verify that the proposed network could meet the requirements of the natural environment, we focus on its detection performance in the areas with uneven pear size and serious background noise. The quantitative test results of SFYOLO with YOLOv5, YOLOv6, and YOLOv7 is shown in Fig. 4. In the comparison of columns a and b, the detection performance of the networks for dense areas containing a large number of small pears was mainly concerned. The results showed that the cases of missed detection and false detection of SFYOLO were less than those of YOLOv6 and YOLOv7, and obviously

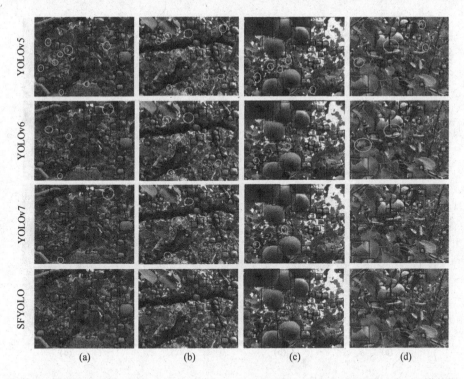

Fig. 4. Comparison of pear detection on the ground dataset, with YOLOv5s, YOLOv6, YOLOv7 and SFYOLO. The cases of missed detection are highlighted in yellow (Color figure online)

less than YOLOv5. This showed that SFYOLO could effectively detect small pears, and was better than YOLOv5, YOLOv6 and YOLOv7 qualitatively. In the comparison of columns of c and d, the detection performance of the networks for areas containing pears with different scales and blocking each other was mainly concerned. The results showed that SFYOLO still had a beneficial effect in detecting pears occluded by leaves or by each other, and the effect was slightly better than that of YOLOv6 and YOLOv7, and obviously better than that of YOLOv5. In summary, SFYOLO could effectively detect pears in the natural environment, especially for small and occluded ones. This helped to provide a lightweight vision system for portable devices and provided auxiliary information for subsequent decision-making such as picking path planning.

5.5   Qualitative Experimental Results on Aerial Dataset

In agricultural monitoring, three-dimensional monitoring was of great significance to promote orchard automation. The aerial images obtained by aircraft could be used for pear growth monitoring, intelligent yield estimation and orchard spraying. In order to verify the performance of the proposed network in

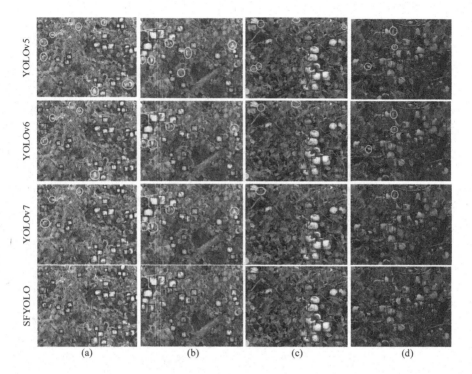

Fig. 5. Comparison of pear detection in the aerial dataset captured by drones from the bird's-eye viewpoint and side viewpoint. The cases of missed detection are highlighted in yellow (Color figure online)

cross-domain detection tasks, aerial dataset including bird's-eye viewpoint and side viewpoint were used to further explore the robustness of the network to the change of viewpoints. Figure 5 shows the qualitative test results of YOLOv5, YOLOv6, YOLOv7, and SFYOLO on the aerial dataset using the weights of network trained on the ground dataset. Columns a and b were images taken by drones from the bird's-eye viewpoint, and columns c and d were images taken by drones from side viewpoint. These images contain a large number of small and occluded pears, which made it much more difficult to detect. Under the influence of light and background noise, the pears in the bird's-eye viewpoint were closer to the color of leaves and soil, so they were more difficult to detect. As could be seen from the figure, the detection performance of SFYOLO was slightly better than that of YOLOv6 and YOLOv7 with less cases of missed detection. Among them, the improvement in the bird's-eye viewpoint was more obvious, indicating that the SFYOLO can effectively reduce the interference of background noise, so as to identify the target more accurately. Compared with the detection ability of YOLOv5, the detection ability of SFYOLO was significantly improved, and the cases of missed detection were greatly reduced in both viewpoints. This benefited from the construction of SFA mechanism, the embedding of space-friendly transformer encoder and the employment of decoupling head, which not only improved the ability of local and global feature extraction, but also enhanced the transfer and generalization ability of the network.

To sum up the above results, due to the effective design of network and the novel improvement of structure, SFYOLO was endowed with sufficient accuracy and robustness improvement, and could carry out cross-domain detection efficiently and accurately. Additionally, SFYOLO could be directly transferred for pear detection in aerial images, which not only verified the generalization ability, but also reduced the workload, improved the production efficiency, bringing potential improvement of production benefit.

5.6 Ablation Studies

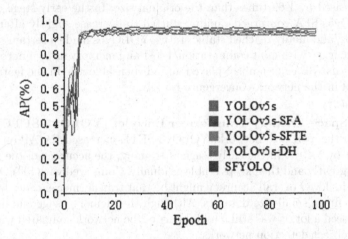

Fig. 6. Line graph of AP metrics for ablation experiments on the ground dataset

To further explore the impact of the improvements of the original YOLOv5s, three sets of ablation experiments were designed in this paper, and the improvement of each improvement on the network performance was discussed based on the results of AP metrics. In order to explore the influence of SFA mechanism on the network, it was added to P3, P4 and P5 in the neck of the original YOLOv5s, which was named as YOLOv5s-SFA. In order to explore the influence of the position of SF-TE in the network, it was applied at the end of backbone, which was named as YOLOv5s-SFTE. In order to explore the impact of decoupled head on performance, decoupled head were applied as head, which was named as YOLOv5s-DH. The results of the ablation experiment are shown in Fig. 6 and Table 2. Each module could contribute to the performance improvement in pear detection on the ground dataset.

Table 2. Results in the ablation experiments.

Models	AP(%)	Detection Time(ms)	Memory usage(MB)
YOLOv5s	91.09	11.2	13.7
YOLOv5s-SFA	91.96(+0.87)	11.8(+0.6)	23.2(+9.5)
YOLOv5s-SFTE	92.44(+1.49)	15.7(+4.5)	113.9(+100.2)
YOLOv5-DH	92.12(+1.03)	12.1(+0.9)	17.9(+4.2)
SFYOLO	93.12(+2.03)	13.2(+2.0)	54.5(+40.8)

Effect of Space-Friendly Attention Mechanism (YOLOv5s-SFA). The AP of YOLOv5s-SFA on the validation set increased by 0.87%. The memory usage increased by 1.69 times than the original size. In the early stage of training, YOLOv5s-SFA converged rapidly, started to decrease slightly after twenty generations, and finally reached stable status. YOLOv5s made a certain improvement in accuracy with faint computational cost and memory usage improvement, indicating that SFA mechanism played an indispensable role in the feature map and helped in the network convergence rapidly.

Effect of Space-Friendly Transformer Encoder (YOLOv5-SFTE). Compared with the YOLOv5 network, YOLOv5-SFTE increased the AP on the validation set by 1.49%. In the early stage of training, the accuracy of the network fluctuates greatly and then keeps stable gradually. Compared with SFYOLO, the reason for its lower overall accuracy might be that transformer encoder was more difficult to fit on small-size datasets. Although the memory usage and detection time increased a lot, it was still a useful part of the network compared with other one-stage object detection networks.

Effect of Decoupled Head (YOLOv5s-DH). Compared with the YOLOv5 network, YOLOv5s-DHhad 1.03% higher AP on the verification set. In the early stage of training, the accuracy of the network fluctuated greatly and then gradually kept relatively stable. Compared with YOLOv5s, the reason for its higher overall accuracy might be that the decoupled anchor-free detectors had almost no manual preset hyperparameters, which could be well migrated to the application scene in this paper to satisfy the object detection requirements. While SFYOLO-DH has achieved effective accuracy improvement, there was almost no significant improvement in detection time and memory usage, which was very suitable for completing high-precision detection tasks in an environment with limited computing resources.

6   Conclusion

Improving the detection ability of the network for target fruits was the basis of implementing automated agronomic management. To address the problem of target fruit detection in real complex scenes, this paper proposed SFYOLO with the inherent characteristics of the aggregation attention of space domain and channel domain. The construction of the Space-Friendly Attention mechanism and the modification of the original transformer encoder enabled the network to aggregate spatial feature and channel feature as well as capture both local and global information in an effective and efficient way. The decoupled anchor-free head enabled the network to complete the tasks of classification, location and regression independently and better, which further enhanced the detection performance. The experimental results showed that our network enhances the feature extraction ability and improved the detection accuracy of small pears and occluded pears. The proposed SFYOLO achieved an average accuracy of 93.12% on the ground dataset and outperformed the typical or recent members of the YOLO series such as YOLOv6 and YOLOv7 on both the ground dataset and the aerial dataset in terms of speed and accuracy. In the future, we will further investigate the ways of implementing pear detection at different growth stages, as well as the reduction of the training and detection cost to better support real-time detection of other fruits.

References

1. Chen, J., et al.: Detecting ripe fruits under natural occlusion and illumination conditions. Comput. Electron. Agric. **190**, 106450 (2021)
2. Perez-Borrero, I., Marin-Santos, D., Gegundez-Arias, M.E., Cortes-Ancos, E.: A fast and accurate deep learning method for strawberry instance segmentation. Comput. Electron. Agric. **178**, 105736 (2020)
3. Sharma, B.B., Kumar, N.: IoT-based intelligent irrigation system for paddy crop using an internet-controlled water pump. Int. J. Agric. Environ. Inf. Syst. (IJAEIS) **12**(1), 21–36 (2021)
4. Sun, Z., Feng, W., Jin, J., Lei, Q., Gui, G., Wang, W.: Intelligent fertilization system based on image recognition. In: 2021 IEEE 6th International Conference on Computer and Communication Systems (ICCCS), pp. 393–399. IEEE (2021)
5. Gai, R., Chen, N., Yuan, H.: A detection algorithm for cherry fruits based on the improved yolo-v4 model. Neural Comput. Appl., 1–12 (2021). https://doi.org/10.1007/s00521-021-06029-z
6. Hu, X., et al.: Real-time detection of uneaten feed pellets in underwater images for aquaculture using an improved YOLO-V4 network. Comput. Electron. Agric. **185**, 106135 (2021)
7. Redmon, J., Farhadi, A.: YOLO9000: better, faster, stronger. In: Proceedings of the IEEE Conference on Computer Vision and Pattern Recognition, pp. 7263–7271 (2017)
8. Redmon, J., Farhadi, A.: YOLOv3: an incremental improvement. arXiv preprint arXiv:1804.02767 (2018)
9. Bochkovskiy, A., Wang, C.-Y., Liao, H.-Y.M.: YOLOv4: optimal speed and accuracy of object detection. arXiv preprint arXiv:2004.10934 (2020)
10. Ge, Z., Liu, S., Wang, F., Li, Z., Sun, J.: YOLOX: exceeding YOLO series in 2021. arXiv preprint arXiv:2107.08430 (2021)
11. Glenn Jocher, A.S.: "YOLOv5." https://github.com/ultralytics/yolov5

12. Hu, J., Shen, L., Sun, G.: Squeeze-and-excitation networks. In: Proceedings of the IEEE Conference on Computer Vision and Pattern Recognition, pp. 7132–7141 (2018)

13. Woo, S., Park, J., Lee, J.-Y., Kweon, I.S.: CBAM: convolutional block attention module. In: Ferrari, V., Hebert, M., Sminchisescu, C., Weiss, Y. (eds.) ECCV 2018. LNCS, vol. 11211, pp. 3–19. Springer, Cham (2018). https://doi.org/10.1007/978-3-030-01234-2_1

14. Hou, Q., Zhou, D., Feng, J.: Coordinate attention for efficient mobile network design. In: Proceedings of the IEEE/CVF Conference on Computer Vision and Pattern Recognition, pp. 13713–13722 (2021)

15. Dosovitskiy, A., et al.: An image is worth 16x16 words: transformers for image recognition at scale. arXiv preprint arXiv:2010.11929 (2020)

16. Carion, N., Massa, F., Synnaeve, G., Usunier, N., Kirillov, A., Zagoruyko, S.: End-to-end object detection with transformers. In: Vedaldi, A., Bischof, H., Brox, T., Frahm, J.-M. (eds.) ECCV 2020. LNCS, vol. 12346, pp. 213–229. Springer, Cham (2020). https://doi.org/10.1007/978-3-030-58452-8_13

17. Graham, B., et al.: Levit: a vision transformer in convnet's clothing for faster inference. In: Proceedings of the IEEE/CVF International Conference on Computer Vision, pp. 12259–12269 (2021)

18. Wu, H., et al.: CvT: introducing convolutions to vision transformers. In: Proceedings of the IEEE/CVF International Conference on Computer Vision, pp. 22–31 (2021)

19. Chen, Z., Xie, L., Niu, J., Liu, X., Wei, L., Tian, Q.: Visformer: the vision-friendly transformer. In: Proceedings of the IEEE/CVF International Conference on Computer Vision, pp. 589–598 (2021)

20. Tian, Z., Shen, C., Chen, H., He, T.: Fcos: fully convolutional one-stage object detection. In: Proceedings of the IEEE/CVF International Conference on Computer Vision, pp. 9627–9636 (2019)

21. Liu, W., et al.: SSD: single shot multibox detector. In: Leibe, B., Matas, J., Sebe, N., Welling, M. (eds.) ECCV 2016. LNCS, vol. 9905, pp. 21–37. Springer, Cham (2016). https://doi.org/10.1007/978-3-319-46448-0_2

22. Li, C., et al.: YOLOv6: a single-stage object detection framework for industrial applications. arXiv preprint arXiv:2209.02976 (2022)

23. Wang, C.-Y., Bochkovskiy, A., Liao, H.-Y.M.: YOLOv7: trainable bag-of-freebies sets new state-of-the-art for real-time object detectors. arXiv preprint arXiv:2207.02696 (2022)

24. Lin, T.Y., et al.: Microsoft COCO: common objects in context. In: Fleet, D., Pajdla, T., Schiele, B., Tuytelaars, T. (eds.) ECCV 2014. LNCS, vol. 8693, pp. 740–755. Springer, Cham (2014). https://doi.org/10.1007/978-3-319-10602-1_48

25. Zhang, H., Cisse, M., Dauphin, Y.N., Lopez-Paz, D.: mixup: Beyond empirical risk minimization. arXiv preprint arXiv:1710.09412 (2017)

26. Song, G., Liu, Y., Wang, X.: Revisiting the sibling head in object detector. In: Proceedings of the IEEE/CVF Conference on Computer Vision and Pattern Recognition, pp. 11563–11572 (2020)

Feature Fusion Expression Recognition Algorithm Based on DCA Dimensionality Reduction

Xiaofang Wang[1](✉)(ID), Dengjie Fang[2], Qianying Zou[1](ID), and Yifei Shi[1]

[1] Geely University of China, Chengdu 641423, Sichuan, China
939549393@qq.com
[2] Chengdu College of University of Electronic Science and Technology of China,
Chengdu 611731, Sichuan, China

Abstract. Aiming at the insufficiency of HOG and LBP feature fusion and imbalance of data samples, a feature fusion algorithm based on DCA dimension reduction is proposed. The algorithm firstly extracts the expression features of the preprocessed data by using the improved HOG-LBP algorithm, then uses the improved discriminative correlation analysis algorithm to reduce the dimension of the feature information. Experimental results show that the improved algorithm can effectively reduce feature dimensions and enhance the relevance of classification data to the largest extent. Compared with several traditional machine learning fusion algorithms, the recognition rate increases by 1.1% on average, the training time decreases by 101.6% on average, and the reasoning speed increases by 74.7%. Although the average recognition rate of the improved algorithm has no advantage over several deep learning algorithms, its training time and reasoning speed are far below the deep learning framework.

Keywords: Discriminant correlation analysis · Feature fusion · Expression recognition · Partial binary tree dual support vector machine multiple classification

Expression is the basis of human feelings and is an important means of communication and communication between people [7]. In recent years, facial expression recognition has become a new direction of computer vision and college psychological education and has become the focus of experts and scholars. It includes two directions: face recognition based on machine learning and face recognition based on deep learning. HOG and LBP are the main research fields of facial expression recognition based on machine learning. Wu Hao [4] and others put forward a feature recognition algorithm combining the DCLBP operator and HOAG operator. The algorithm uses DCLBP and HOGA to extract local texture features and local contour features respectively, then uses CCA to fuse features, and then uses SVM to classify. Although the method can achieve a good recognition rate, it has a poor processing effect in unbalanced datasets and complex large calculations of data dimensions. Qiao Ruiping [14] and others proposed an improved

HOG-LBP multi-classification algorithm. The algorithm extracts the features of the fused HOG-LBP operator and then uses the combination method to correct the features. Finally, the binomial tree support vector machine is used to classify the targets. The method can best achieve multi-classification, but the calculation dimension is relatively high. Huang Xinghua [19] et al put forward an image recognition and extraction algorithm for texture feature fusion. This algorithm uses AD-LBP and HOG to extract the image and then uses SVM to classify the image. Although this algorithm can effectively identify the image content centroid, the feature dimension is higher, and the model calculation is complex. Face recognition based on deep learning is mainly focused on CNN, AlexNet, VGG network, and so on. Among them, WALECKI [1] and others put forward the CNN architecture for end-to-end learning. Although the model can well reduce the dependence on the face model and preprocessing, the training time of the model is long, and the requirement on hardware resources is high; Sun Rui [3] and others proposed the rotation invariant face detection of cascade network and pyramid optical flow. Although the algorithm has a better effect on video flow processing and jitter processing, it has higher model complexity, longer model training time, and higher demand on hardware resources; Yang Gran [15] and others proposed the face recognition model based on the convolution neural network in time and space domain, which can effectively reduce the dependence on manual design features and has better adaptability, but the algorithm architecture is complex, and the model training time and reasoning speed are slow. Based on the above problems, this paper proposes a feature fusion facial expression recognition algorithm based on discriminant correlation analysis to realize the classification of unbalanced samples. The main contributions of the algorithm are:

1. to obtain better facial texture and improve the recognition rate, the improved MB-LBP algorithm is used to extract texture features.
2. to retain feature information better, reduce model training time and detection time, the DCA algorithm fusing PCA dimensionality reduction is used to reduce contour features and texture features and fuse the real world.

1   Introduction to Relevant Technologies

The Local Binary Patterns (LBP) [17] algorithm is one of the most classical feature extraction algorithms in 1999. Based on LBP, MB-LBP was proposed by Shengcai Lia and others in 2007 to reduce the binary mode of the traditional LBP algorithm, realizing the reduction of the traditional binary mode from 256 to 59.

Discriminant Correlation Analysis [10] (DCA), first proposed by Haghighat et al in 2016, is used for feature texture recognition and fusion, which combines canonical correlation analysis [5,6] and linear discriminant analysis, minimizing the correlation between different categories of data while maximizing the two types of data.

2   Algorithmic Implementation

2.1   Overall Architecture

The implementation of the algorithm is divided into three stages: image pre-processing, feature extraction and fusion and expression classification. The algorithm flow is shown in Fig. 1.

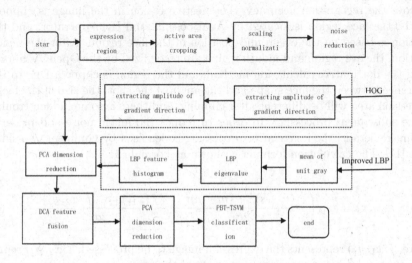

Fig. 1. Algorithm flow chart

As shown in Fig. 1, the implementation steps of the algorithm presented in this article are:

1. AdaBoost combined with the Haar operator locates the human face region, clips the images based on the eye and mouth regions, and unifies the image size by image normalization;
2. Using the bilateral filtering algorithm to reduce the noise of the image to reduce the interference of the image processing;
3. Using the HOG algorithm to calculate the gradient value of the image to obtain the expression contour features;
4. Using the improved LBP algorithm to divide the feature image and obtain the expression texture feature;
5. Using PCA to reduce the dimension of the contour feature and texture feature, so that the contour feature and texture feature dimension are consistent for feature fusion;

6. Normalized contour features and texture features, and use DCA to fuse the two groups of features to obtain the fusion expression features;
7. PCA is used again to reduce the computational complexity of facial expression discrimination and reduce the training time;
8. The partial binary tree dual support vector machine multi-classification algorithm is used to classify the expression features of the unbalanced dataset, and the final result of expression discrimination is obtained.

2.2   Implementation Process

Image Preprocessing. To eliminate the redundant image information and improve the recognition accuracy, the effective region in the image is clipped. Firstly, the face region is located by AdaBoost [9] and Haar operator, and the location is based on the eyeground position of the face region. After the region location, the eye region and mouth region are recognized by the OpenCV library using the dot matrix calculation method, and the region is clipped. Due to the difference between the face shape and the angle of the shot, the size of the facial expression area will be different after clipping, and there is some hidden trouble in the subsequent processing. In order to reduce the image noise and preserve the image details, the image after processing is denoised by nonlinear two-sided filter [11]. The calculation formula is shown as Formula 1.

$$w(n, m, s, r) = \exp\left(-\frac{(n-s)^2 + (m-r)^2}{2\sigma^2} - \frac{\|f(x_1, y_1) - f(x_2, y_2)\|^2}{2\sigma^2}\right) \quad (1)$$

where, $f(x_1, y_1)$ represents the expression image to be processed, $f(x_2, y_2)$ represents the pixel value of the expression image at the point (x_2, y_2), (s, r) represents the central coordinates of the template image, (n, m) represents the coordinates of other coefficients of the expression template image, and σ represents the standard deviation information of the Gaussian function, i.e. the influence factor of denoising processing.

After pruning and denoising, we get the expression data set, which is shown in Fig. 2.

(a) Original image   (b) Image lattice clipping   (c) Normalization and enhancement

Fig. 2. Image preprocessing stage diagram

Feature Extraction. In order to reduce the inefficiency of feature extraction caused by local shadows, this paper uses Gamma correction [23] to adjust the image contrast to improve the sharpness of HOG algorithm. The formula for Gamma correction is shown in Formula 2.

$$f(I) = I^\gamma \tag{2}$$

where, I is the expression image to be corrected, γ is the correction factor, and $\gamma = 0.5$ is the best. To get the expression contour feature, this paper uses the HOG algorithm to extract the contour feature vector [8,22] from the corrected expression image and calculates the gradient in the horizontal and vertical direction of the image respectively to obtain the amplitude information of the contour pixels. The calculation formula is shown in Formula 3.

$$\nabla f(x,y) = \begin{bmatrix} g_x \\ g_y \end{bmatrix} = \begin{bmatrix} \frac{\partial f}{\partial x} \\ \frac{\partial f}{\partial y} \end{bmatrix} = \begin{bmatrix} f(x+1,y) - f(x-1,y) \\ f(x,y+1) - f(x,y-1) \end{bmatrix} \tag{3}$$

where, g_x and g_y are the gradient values in the horizontal and vertical directions of pixels (x,y), respectively. The gradient amplitudes $g(x,y)$ and gradient angles α of pixels (x,y) are calculated, where the gradient amplitudes and angles are calculated as shown in Formula 4 and Formula 5.

$$g(x,y) = \sqrt{g_x^2 + g_y^2} \tag{4}$$

$$\alpha = \arctan\left(\frac{g_y}{g_x}\right) \tag{5}$$

In this paper, the gradient direction is scaled from 0 to 180°. After gradient calculation, the image is divided into innumerable 8 × 8 pixel cell units, each of which is composed of 2 × 2 cell units. To make the feature fusion more convenient, the paper divides each cell into 15 bins, and sets the angle of each bin to 12°, counts the gradient angle of each pixel in each cell, and adds up the corresponding gradient amplitude of each bin, which is the histogram of each cell.

After getting the gradient histogram of the image composed of cell units, the cell units in each block are concatenated, and the histogram features of each block are merged to form the HOG feature vectors of the image. The calculation formula is shown in Formula 6.

$$I' = \frac{I - \min}{\max - \min} \tag{6}$$

where, I represents a number in the pixel data of the expression image, min represents the smallest value in the pixel data, max represents the largest value in the pixel data, and I' represents the outline of the expression image. To get more details of expression image texture and get better expression target texture

(a) UP-LBP

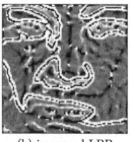

(b) improved LBP

Fig. 3. UP-LBP and improved LBP processing results

features, this paper proposes an improved LBP algorithm for feature extraction, UP_LBP, and improved LBP. The result as shown in Fig. 3.

Based on the original LBP algorithm, the improved LBP algorithm uses MB-LBP [18] to optimize the feature data. The expression image is divided into pixels as the unit of image block operation, and the average gray value of the expression image is calculated to classify and extract the texture features.

After optimizing the feature data of facial expression images by MB-LBP, we use UP-LBP [22] to extract the specific facial expression texture features. Texture feature extraction takes each pixel of the expression image as the center pixel, compares it with the gray value of 8 neighboring pixels, and then combines it clockwise or counterclockwise to get a set of binary codes corresponding to the center pixel. The calculation formula is shown in Formula 7 and Formula 8.

$$S(m) = \begin{cases} 1, x \geq 0 \\ 0, x < 0 \end{cases} \tag{7}$$

$$LBP(c) = \sum_{i=0}^{P-1} S(x_i - x_c) 2^i \tag{8}$$

where, P represents the number of pixels of adjacent points in the expression image, and x_i represents the gray value of any of the 8 adjacent pixels in the expression image.

The resulting central pixel is judged in binary encoded form. When there is a $0 \longrightarrow 1$ or $1 \longrightarrow 0$ transform in the number of cyclic binary encodings, the number of hops ≤ 2 is labeled as an equivalent pattern class, and the corresponding decimal result is the LBP value. The rest of the non-equivalent pattern coding is classified as a mixed pattern class, and the encoding value is set to 0.

After coding and judging, the LBP value of the whole image is obtained, and its value is transformed into an image feature vector. To fuse the features and reduce the differences, the image is divided into cell units of the same scale as the HOG algorithm. In this paper, the maximum gray value of the LBP feature value is taken as the upper bound, and the interval (0, LBP(max)) is divided into 15 equal parts to count the features of each cell unit and form a feature histogram.

In the improved LBP algorithm, the cell units are concatenated with HOG algorithm to get the histogram of a block, then all the blocks are concatenated to get the LBP feature vectors of the image, and the texture feature of the expression image is obtained by normalization.

After being processed by HOG and the improved LBP algorithm, the feature vectors of expression contour and texture are obtained, and the two sets of feature vectors have different dimensions.

Feature Dimension Reduction and Fusion. To fuse the contour features and texture features of different dimensions and reduce the training time of expression recognition effectively, the PCA algorithm [12,13,20] is used to reduce the dimension of two sets of feature vectors. In this paper, the dimension of each group of a characteristic matrix is reduced, and then all samples are centralized as shown in Formula 9.

$$\sum_{i=1}^{m} x^{(i)} = 0 \qquad (9)$$

where, $x^{(i)}$ represents any n-dimensional data in the original eigenvector data matrix, and m represents the number of the expression data matrix. After centralization processing, the expression feature samples are transformed by projection, and the processing results are shown in Formula 10.

$$\left\{ x^{(1)}, x^{(2)}, \ldots, x^{(m)} \right\} \rightarrow \{\omega_1, \omega_2, \ldots, \omega_d\} \qquad (10)$$

where, $d \leq m$, ω represents the standard orthogonal basis for dimensionality reduction, and its value is calculated from Formula 11.

$$\begin{cases} \omega_i^T \omega_j = 0 \\ \|w\|_2 = 1 \end{cases} \qquad (11)$$

After the projection transformation processing, the sample point $x^{(i)}$ of the expression feature is projected to the D dimension hyperplane as shown in Formula 12.

$$\xi^{(i)} = \left(\xi_1^{(i)}, \xi_2^{(i)}, \ldots, \xi_d^{(i)} \right)^T \qquad (12)$$

where, $\xi_j^{(i)}$ represents the expression $x^{(i)}$ projected on the jth dimension of d, which is calculated as Formula 13.

$$\xi_j^{(i)} = \omega_j^T x^{(i)} \qquad (13)$$

After processing, the expression feature vector of the hyperplane is used to deduce the dimensionality of the data, and the formula is shown in Formula 14.

$$\overline{x}^{(i)} = \sum_{j=1}^{d} \xi_j^{(i)} \omega_j \qquad (14)$$

To order to make all the feature vector data samples in the feature matrix nearest to the hyperplane and reduce the inaccuracy of recognition caused by dimensionality reduction, the distance calculation formula is used to obtain the distance, which is shown in the Formula 15.

$$Distance = \sum_{i=1}^{m} \left\| \bar{x}^{(i)} - x^{(i)} \right\|_2^2 \tag{15}$$

Since the expression information is two sets of feature vectors, to facilitate the fusion of the feature matrix, the Formula 15 is transformed into a feature matrix, as shown in the Formula 16.

$$Distance = -tr\left(W^T X X^T W\right) + \sum_{i=1}^{m} \left(x^{(i)}\right)^T x^{(i)} \tag{16}$$

where, $\sum_{i=1}^{m} \left(x^{(i)}\right)^T x^{(i)}$ represents the covariance matrix of the expression feature vector data set, and W represents the dimensionality reduction processing transformation matrix. The transformation matrix corresponding to the minimum Distance of the projection is composed of the eigenvectors corresponding to the first d eigenvalues in the covariance matrix XX^T.

After processing, the result is transformed into the eigenvector matrix using the eigenmatrix transformation formula, and the calculation is shown in Formula 17.

$$z^{(i)} = w^T x^{(i)} \tag{17}$$

After processing, the original expression feature samples are transformed into the corresponding result of the minimum projection distance, which is the reduced dimension expression feature.

After dimensionality reduction by PCA, this paper uses the DCA algorithm [21] to fuse the two sets of feature vectors, so as to get the fused feature information for expression discrimination. DCA feature fusion calculates the average value of each dataset in the whole feature data set, then calculates the relationship between the facial features, reduces the dimension, and fuses the features. The formula for calculating the average value of each dataset is shown in Formula 18.

$$\bar{x}_i = \frac{1}{n_i} \sum_{j=1}^{n} x_{ij} \tag{18}$$

where, $\bar{x}_i$ represents the average value of the data set for Class i, and n represents the number of emoticons under Class i. After getting the average of each kind of dataset through the average calculation formula, the average calculation formula is used again to get the average value of the whole feature dataset, as shown in Formula 19.

$$\bar{x} = \frac{1}{n_i} \sum_{i=1}^{c} n_i \bar{x}_i \tag{19}$$

where, c represents the number of sets of features in an expression image, and there are 7 types of expressions in the text, i.e. c = 7, $\bar{x}$ represents the average of the whole feature set.

To compute the relationship between expression features, this paper uses a divergence matrix to compute the relationship between expression classes. The formula is shown in Formula 20.

$$S_{bx(p \times q)} = \Phi_{bx} \Phi_{bx}^T \tag{20}$$

where, Φ_{bx} the calculation is shown in Formula 21.

$$\Phi_{bx(p \times c)} = \left[\sqrt{n_1} \left(\bar{x}_1 - \bar{x} \right), \ \sqrt{n_2} \left(\bar{x}_2 - \bar{x} \right), \ \ldots, \ \sqrt{n_c} \left(\bar{x}_c - \bar{x} \right) \right] \tag{21}$$

To get a more accurate classification, this paper optimizes the divergence matrix and transforms the divergence matrix into a diagonal matrix. The formula is shown in Formula 22.

$$P^T \left(\Phi_{bx}^T \Phi_{bx} \right) P = \vec{\Lambda} \tag{22}$$

where, P represents the orthogonal eigenvector matrix, and $\vec{\Lambda}$ represents the diagonal matrix of the real number of non-negative eigenvalues. Convert from Formula 22 to Formula 23, as shown in Formula 23.

$$\varphi^T \left(\Phi_{bx}^T \Phi_{bx} \right) \varphi = \Lambda_{(r \times r)} \tag{23}$$

Then, the effective eigenvectors of S_{bx} are obtained by calculating the mapping. The formula is shown in Formula 24.

$$\left(\Phi_{bx} \varphi \right)^T S_{bx} \left(\Phi_{bx} \varphi \right) = \Lambda_{(r \times r)} \tag{24}$$

By transforming Formula 24, the dimension of the characteristic image X is reduced from p to r. where, $w_{bx} = \Phi_{bx} \varphi \Lambda^{-\frac{1}{2}}$, after transformation, as shown in Formula 25.

$$\begin{cases} W_{bx}^T S_{bx} W_{bx} = I \\ X'_{(r \times n)} = W_{bx(r \times p)}^T X_{(p \times n)} \end{cases} \tag{25}$$

Similarly, do the same transformation to the characteristic matrix Y, get Y', and then construct the cross-covariance matrix for the transformed two groups of characteristic data, as shown in Formula 26.

$$S'_{xy} = X' Y'^T \tag{26}$$

Then the cross-covariance matrix is decomposed into the singular values, and the singular value decomposition theorem is used to compute the singular values. The formula is shown in Formula 27.

$$S'_{xy} = U \sum V^T \tag{27}$$

where, $\sum$ is calculated by Formula 28.

$$\sum = U^T S'_{xy} V \tag{28}$$

The left and right singular vectors of Formula 28 are $a = U\Sigma^{-\frac{1}{2}}$ and $b = V\Sigma^{-\frac{1}{2}}$ respectively, and the Formula 28 is converted to Formula 29.

$$I = \left(U\Sigma^{-\frac{1}{2}}\right)^T S'_{xy} \left(V\Sigma^{-\frac{1}{2}}\right) \tag{29}$$

Finally, the feature data set is transformed into the corresponding feature vector matrix in the X and Y directions, and the calculation is shown in formula Formula 30.

$$X^* = a^T X'$$
$$Y^* = b^T Y' \tag{30}$$

The final fusion eigenvectors are calculated by using the summation formula for X and Y feature matrices. The formula is shown in Formula 31.

$$Z = X^* + Y^* = \begin{pmatrix} W_X \\ W_Y \end{pmatrix}^T \begin{pmatrix} X \\ Y \end{pmatrix} \tag{31}$$

where, Z is the fusion result of the facial features sought.

PBT-TSVM Emoticons Classification. In order to shorten the training time and get the accurate classification of facial expression feature, the PCA is used to reduce the dimension of the fusion feature, and the PBT-TSVM [2] algorithm is used to classify the feature after reducing the dimension. First, K class (7 kinds) expression feature problems are transformed into K-1 binary classification problems by using partial binary tree structure, and then each two kinds of problems are classified by using double support vector machines.

The first TSVM is constructed, neutral expression is marked as positive, the other is marked as negative, and TSVM1 is trained. Two non-parallel expression superplanes L1 and L2 are obtained. L2 is used to test the expression superplanes until the K -1 TSVM is constructed.

Where, the training of TSVM is to transform the quadratic optimization problem into two smaller QPPs, and obtain two uneven classification hyperplanes. The solutions are shown in Formula 32 and Formula 33.

$$\begin{cases} \min_{p^{(i1)},q^{(i1)}} \frac{1}{2} \left\| K\left(a_i, a'\right) p^{(i1)} + \gamma_{i1} q^{(i1)} \right\|^2 + \beta_{i1} \gamma'_{i2} \delta s.t. \\ \qquad - \left(K\left(\tilde{a}, a'\right) p^{(i1)} + \gamma_{i2} q^{(i1)}\right) + \delta \geq \gamma_{i2} \\ \min_{p^{(i2)},q^{(i2)}} \frac{1}{2} \left\| K\left(a_i, a'\right) p^{(i2)} + \gamma_{i2} q^{(i2)} \right\|^2 + \beta_{i2} \gamma'_{i1} \delta s.t. \\ \qquad - \left(K\left(\tilde{a}, a'\right) p^{(i2)} + \gamma_{i1} q^{(i2)}\right) + \delta \geq \gamma_{i1} \end{cases} \tag{32}$$

where, $\delta \geq 0$,because there are 7 expressions, K = 7, a_i represents the i positive sample, a' represents all samples $i + 1$ to K, γ_{i1} and γ_{i2} represent feature vector units, p and q represent the optimal hyperplane normal vectors and offsets of the expression features, and δ represents relaxation variables. By solving the Formula 32, we obtain an expression category, as shown in the Formula 33.

$$\begin{cases} K\left(a, a'\right) p^{(i1)} + q^{(i1)} = 0 \\ K\left(a, a'\right) p^{(i2)} + q^{(i2)} = 0 \end{cases} \tag{33}$$

If the distance from the positive hyperplane to the negative hyperplane is less than the distance from the positive hyperplane to the negative hyperplane, it is a positive class; otherwise it is a negative class. Thus, the expression category to which the fused feature information belongs is obtained.

3   Analysis of Experimental Results

3.1   Experimental Environment

This paper uses SPYDER IDE to compile and Python to program. The hardware uses an 8-core Intel Core i5-7200 U2.50 GHz processor, 32 GB memory, and 64-bit Windows 10 environment. The environment parameters are shown in Table 1.

Table 1. Test environment parameters

Experimental tool	Programming languages	Platform	CPU models	Memory	OSs
SPYDER IDE	Python	ThinkPad	Intel Core i5-7200U2.50 GHz	32 GB	64bit Windows 10

This article uses open-source JAFFE data sets for validation. The dataset contains 6 types of facial expression images and 1 type of neutral expression image composed of 10 Japanese women, as shown in Fig. 4. From left to right, Happy, Anger, Sad, Fear, Disgust, Surprise, and Neutral are shown in the JAFFE data set. The data set of neutral, sad and wearisome is augmented randomly to enlarge the sample size and obtain the imbalanced data set of the category sample.

3.2   Experimental Results and Analysis

Analysis of Different Algorithm Results. Under the same experimental environment, this paper adopts the HOG algorithm, UP-LBP algorithm, fusion HOG-LBP algorithm, fusion DCLBP and HOAG [7], improved HOG-LBP multi-classification algorithm [4], AD-LBP HOG algorithm [14] and improved algorithm to carry out 10000 iterative training on data sets, calculates training time,

Fig. 4. Seven sample expressions from the JAFFE database

and selects 7 expressions of 10 sets of sample pictures randomly from the test set after training, calculates the average recognition rate of each group of 14 expressions, and calculates the calculation result with one decimal place, inference time and three decimal places. The calculation result is shown in Table 2.

Table 2. 7 Average recognition rate (%) and training time (s) for different algorithms

Serial number	HOG	UP-LBP	HOG-LBP	DCLBP-HOAG	Improved HOG-LBP	AD-LBP HOG	Improved algorithm
1	79.1	81.7	86.4	91.3	89.2	90.1	**91.5**
2	75.5	77.4	84.5	90.1	87.8	88.3	**89.2**
3	79.6	79.6	87.5	90.7	89.2	89.9	**90.8**
4	78.5	79.7	86.5	91.5	90	90.2	**91.8**
5	79.7	80.6	90.1	92.5	90.1	90.8	**92.7**
6	79.5	79.2	89.8	91.8	89.9	90.5	**92**
7	78.7	78.7	87.2	91	89.2	89.8	**91.2**
8	75.9	77.4	84.8	89.8	87.6	87.2	**89.5**
9	82.2	83.5	90.4	92.9	91.8	92.1	**93.2**
10	83.5	84.2	91.7	94	92.1	92.8	**94.4**
Mean value	79.2	80.2	87.9	91.5	89.7	91.2	**91.6**
Training time	60	65	68	48	62	72	**31**
Inference time	0.065	0.079	0.081	0.056	0.078	0.092	**0.019**

As can be seen from the above table, the average recognition rate of the HOG algorithm, UP-LBP algorithm, fusion of HOG and LBP algorithm, the fusion of DCLBP and HOAG, improved HOG-LBP multi-classification algorithm, AD-LBP HOG algorithm and improved algorithm for JAFFE dataset is 79.2%, 80.2%, 87.9%, 91.5%, 89.7%, 91.2%, and 91.6% respectively. The average recognition rate of the improved algorithm is increased by 15.6%, 14.2%, 4.2%, 0.1%, 0.4%, and 2.1% respectively, compared with that of HOG, UP-LBP, the fusion of HOG and LBP algorithm, the fusion of DCLBP and HOAG, improved HOG-LBP multi-classification algorithm, improved AD-LBP + HOG algorithm. The average recognition rate of the improved algorithm is increased by 6.1% compared with the traditional algorithm. The average training time of HOG algorithm, UP-LBP algorithm, the fusion of HOG and LBP algorithm, the fusion of DCLBP and HOAG algorithm, improved HOG-LBP feature multi-classification algorithm, AD-LBP HOG algorithm, and improved algorithm for processing JAFF dataset is 93.5%, 109.7%, 119.0%, 54.8%, 100.6%, and 132.2% respectively, compared with HOG, UP-LBP, the fusion of Hog and LBP algorithm, the fusion of DCLBP and HOAG, improved HOG-LBP feature multi-classification algorithm, AD-LBP + HOG algorithm, the average training time of the improved algorithm is less than that of 93.5%, 109.7%, 119.0%, 54.8%, 100.6%, and 132.2% respectively. Compared with several traditional algorithms, the training time of the improved algorithm is reduced by 101.6% on average, and the training time of the improved algorithm can be greatly shortened on model training. HOG algorithm, UP-LBP algorithm, the fusion of HOG and LBP algorithm, the fusion of DCLBP and HOAG, improved HOG-LBP feature multi-classification algorithm, AD-LBP HOG algorithm, and improved algorithm for real-time detection of JAFF dataset images, the inference time is 0.065 s, 0.079 s, 0.081 s, 0.056 s, 0.077 s, 0.092 s, and 0.019 s respectively. The average inference time of the improved algorithm is reduced by 0.046 s, 0.06 s, 0.06 s, 0.062 s, 0.037 s, 0.099 s, and 0.019 s respectively, compared with HOG, UP-LBP, fusion Hog, and LBP algorithm, fusion DCLBP, and HOAG algorithm, improved HOG-LBP feature multi-classification algorithm, AD-LBP + HOG algorithm, and improved HOGBP + HOG algorithm, compared with 0.06s, 0.06s, 0.06s, 0.055s, and 0.073s, compared with several traditional algorithms, the average inference time is reduced by 7.7%, and the improved algorithm has some advantages in the application speed of image inference.

Comparison and Analysis of Deep Learning Algorithms. This paper uses CPU to compute VGG19 [16], Deep space-time convolution network [3], STCNN [15], and this algorithm to do 10000 iterative training on the dataset, calculates the training time, and selects 14 pictures of 7 expressions at random for detection, calculates the average recognition rate of the detection results. The results keep one decimal place, and the average inference time of the pictures is shown in Table 3.

Table 3. Comparison and analysis of deep learning algorithms

Algorithmic model	Average recognition rate(%)	Training time(s)	Inference time(s)
VGG19	90.2	>>3600	6.23
Deep space-time convolution network	92.1	>>3600	7.15
STCNN	92.8	>>3600	8.3
Improved algorithm	91.6	31	0.019

As can be seen from the above table, the average recognition rate of VGG19, Deep space-time convolution network, STCNN, and the improved algorithm after processing the JAFFE dataset is 90.2%, 92.1%, 92.8%, and 91.6% respectively, and the average recognition rate of the improved algorithm is 1.5% higher than that of VGG19 network, 0.5% and 1.2% lower than that of Deep space-time convolution network and STCNN network. The model training time of VGG19, Deep space-time convolution network, STCNN network is much higher than 3600 s, the improved algorithm is only 31s, the improved algorithm has a deeper learning framework, the training time is greatly reduced, and the resources consumed are less. In fact, the reasoning time of the improved algorithm is 6.23 s, 7.15 s, 8.30 s 0.019 s, and the reasoning time of the improved algorithm is 6.211 s, 7.131 s, and 8.28 s respectively, compared with the deep learning model. Compared with the deep convolution framework, the improved algorithm has a slightly lower recognition rate, but its training resources and training time are far less than the deep learning framework.

Analysis of Average Recognition Rate of Different Contribution. In the same experimental environment, we reduce the eigenvector to 9 dimensions, and use the HOG algorithm, UP-LBP algorithm, fusing HOG and LBP algorithm, and improved algorithm to deal with them respectively. The Result as shown in Fig. 5.

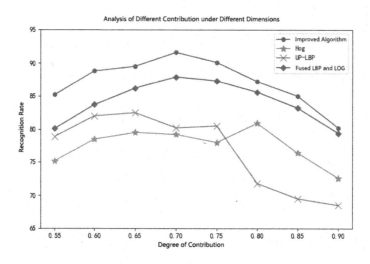

Fig. 5. Contribution analysis of different algorithms in the same dimension

As can be seen from the graph above, the average recognition rate of expression recognition of the HOG algorithm increases from 0.55 to 0.8 on the whole and decreases from 0.8 to 0.95 based on different contribution degrees of the same dimension; the average recognition rate of expression recognition of UP-LBP algorithm increases from 0.55 to 0.65 on the whole and decreases from 0.65 to 0.95 based on different contribution degrees of the same dimension; and the average recognition rate of expression recognition of HOG and LBP algorithm and improved algorithm increase from 0.55 to 0.7 on the whole and decreases from 0.7 to 0.95 based on different contribution degrees of the same dimension. Compared with the HOG algorithm, UP-LBP algorithm, and the fusion of HOG and LBP algorithm, the improved algorithm has the best overall average recognition rate in the same dimension, and the average recognition rate of the four algorithms is between 0.65 and 0.78.

Dimension Average Recognition Rate Analysis. According to the above experiments, in the same experimental environment, the paper selects 0.7 as the contribution rate, processes the dimension of the improved algorithm, reduces the dimension of the feature dimension to 3D, 5D, 7D, 9D, and 11D respectively, and selects 7 expressions of 10 groups of sample pictures at random, and calculates the average recognition rate of 14 samples in each group. The result remains one de. The result as shown in Table 4.

Table 4. Analysis of average recognition rates of algorithms in different dimensions (%)

Dimension	1	2	3	4	5	6	7	8	9	10	Mean value
5 dimensions	82.1	85.6	84.2	88.3	87.2	86.3	84.2	89	87.1	90.4	86.4
7 dimensions	85.6	87.4	86.5	90.1	89.9	87.5	89.6	90.1	89.4	92.5	88.8
9 dimensions	91.5	89.2	90.8	91.8	92.7	92	91.2	89.5	93.2	94.4	91.6
11 dimensions	87.2	88.5	90	89.5	89.5	86.8	91.2	89.4	89.2	90.5	89.2

It can be seen from the above table that the average recognition rate of 10 facial expressions of 5D, 7D, 9D, and 11D is 86.4%, 88.8%, 91.6%, and 89.2% respectively while the contribution rate is 0.72. Under the best condition, the average recognition rate increases from 5D to 9D, and the average recognition rate decreases from 9D to 11D. When the expression feature is 9D, the average recognition rate of the improved algorithm is the best.

Analysis of Results from Different Classifiers. In the same experimental environment, 5 classifiers, KNN, SVM, K-SVM, BT-SVM, and PBT-TSVM, were used to train the processed features for 10000 iterations. After training, 7 expressions of 10 samples were selected randomly, and the average recognition rate and training time of each group were calculated. The results were one decimal place, real-time detection time, and three decimal places. The result as shown in Table 5.

The average recognition rates of KNN, SVM, KSVM, BT-SVM, BT-SVM and PBT-TSVM are 84.1%, 89.7%, 91.1%, 92.0% and 94.4%, respectively. The average recognition rate of the KNN classifier is the lowest, followed by SVM, KSVM, BT-KSVM, and PBT-SVM. The average recognition rate is 12.8%, 5.2%,

Table 5. Recognition rate (%) and training time analysis of different class

Category	1	2	3	4	5	6	7	8	9	10	Training time	Inference time
KNN	81.1	80.9	82.2	82.4	85.1	84.3	82	81.4	83.2	84.1	56	0.031
SVM	86.7	84.1	85.2	85.5	89.8	87.3	86.1	87.4	88.1	89.7	52	**0.026**
KSVM	90.6	88.1	89.4	89.8	90.7	91.4	90	89.3	90.2	91.1	45	**0.02**
BT-SVM	90.8	88.9	89.5	90.2	90	91.6	89.9	89	90.5	92	40	**0.021**
PBT-TSVM	91.5	89.2	90.8	91.8	92.7	92	91.2	89.5	93.2	94.4	31	**0.019**

3.6% and 2.5%, respectively, compared to ST-SVM. The training time of KNN, SVM, KSVM, BT-SVMKNN, and BT-SVMKNN is 56 s, 52 s, 45 s, the 40 s, and 31 s, respectively. The training time of KNN is the longest, followed by SVM, KSVM, and BT-SVM. The shortest training time of PBT-TSVM is 44.6% lower than that of KNN, 40.4% lower than that of SVM, 31.1% lower than that of KSVM, and 25.8% lower than that of BT-SVM, and 35.47% lower than that of PBT-TSVM. Compared with KNN, SVM, KSVM, BT-SVMKNN and BT-SVMKNN classifiers, the inference time of PBT-TSVM is 0.031 s, 0.026 s, 0.020 s, 0.021 s and 0.019 s respectively. Compared with KNN, SVM, KSVM, and BT-SVM, the inference time of PBT-TSVM decreases by 0.012 s, 0.007 s, 0.001 s, and 0.002 s, and the average inference speed increases by 20.0%. PBT-TSVM classifiers have some advantages in recognition rate, model training time, and inference speed.

4   Conclusion

This paper uses HOG and improved LBP to extract contour features and texture features respectively, and uses PCA to reduce the dimension of the expression vector twice, uses DCA to fuse the reduced dimension features, and finally uses PBT-TSVM to classify expression features. This algorithm can be used not only in the field of expression recognition but also in the field of vehicle and traffic target detection and recognition. In this paper, the partial binary tree is used to classify SVM, which can improve the recognition rate of the unbalanced dataset and reduce the computational complexity and training time. Although this paper has some advantages in training time and reasoning speed in dealing with imbalanced datasets and datasets, it still needs to be improved on multi-target discrimination.

References

1. Deep Structured Learning for Facial Action Unit Intensity Estimation. IEEE, Honolulu, HI (2017)
2. Wu, E., Lu, J.: Research on multiclass classification algorithms based on binary tree support vector machines. J. Chongqing Norm. Univ. (Nat. Sci. Edn.) **33**(03), 102–106 (2016)

3. Yang, G., Deng, X., Liu, X.: Facial expression recognition model based on deep time-space convolution neural network. J. Cent. South Univ. (Nat. Sci. Edn.) **47**(7), 2311–2319 (2016)
4. Wu, H., Hu, M., Gao, Y., et al.: Facial expression recognition combining DCLBP and Hoag features. Electron. Meas. Instr. **34**(02), 73–79 (2020). https://doi.org/10.13382/j.jemi.B1901968 https://doi.org/10.13382/j.jemi.B1901968 https://doi.org/10.13382/j.jemi.B1901968
5. Zhang, H., Chen, X.H., Lan, Y.: Supervised canonical correlation analysis based on deep learning. Data Acquisit. Process. 1–8 (2021). https://doi.org/10.19678/j.issn.1000-3428.0061291 https://doi.org/10.19678/j.issn.1000-3428.0061291
6. Zhang, E.H., Chen, X.H.: Low resolution face recognition algorithm based on consistent discriminant correlation analysis. Data Acquisit. Process. **35**(06), 1163–1173 (2020). https://doi.org/10.16337/j.1004-9037.2020.06.016 https://doi.org/10.16337/j.1004-9037.2020.06.016
7. Ye, J., Zhu, J., Jiang, A.V., et al.: Summary of facial expression recognition. Data Acquisit. Process. **35**(01), 21–34 (2020). https://doi.org/10.16337/j.1004-9037.2020.01.002
8. Cheng, K., Shi, W., Zhou, B., et al.: Robust KCF pedestrian tracking method for complex scenes. J. Nanjing Univ. Aeronaut. Astronaut. **51**(05), 625–635 (2019). https://doi.org/10.16356/j.1005-2615.2019.05.007 https://doi.org/10.16356/j.1005-2615.2019.05.007
9. Liang, L., Sheng, X., Lan, Z., et al.: Retinal blood vessel segmentation algorithm based on multiscale filtering. Comput. Appl. Softw. **36**(10), 190–196, 204 (2019)
10. Haghighat, M., Abdel-Mottaleb, E.M., Alhalabi, W.: Discriminant correlation analysis: real-time feature level fusion for multimodal biometric recognition. IEEE Trans. Inf. Forensics Secur. **11**(9), 1984–1996 (2016)
11. Hu, M., Teng, D., Wang, X.H., et al.: Facial expression recognition incorporating local texture and shape features. J. Electron. Inf. **40**(06), 72–78 (2018)
12. Han, P., Wan, Y., Liu, Y., et al.: Airport runway detection with polarimetric SAR image classification by LBP. J. Chin. Graph. **26**(04), 952–960 (2021)
13. Yao, L.P., Pan, Z.L.: Face recognition based on improved hog and LBP algorithms. Optoelectronics **40**(02), 114–118, 124 (2020)
14. Qiao, R., Wang, F., Dong, R., et al.: Multi objective classification simulation based on improved Hog-LBP feature. Comput. Simul. **37**(11), 138–141 (2020)
15. Sun, R., Kan, J., Wu, L., et al.: Rotational invariant face detection in cascade networks and pyramid optical flow. Optoelectron. Eng. **47**(1), 22–30 (2020)
16. Wei, W., Meng, F., Cai, Z., et al.: Summary of facial expression recognition. J. Shanghai Electr. Power Univ. **37**(06), 597–602 (2021)
17. Xiong, W., Yang, X., et al.: Wood defect identification based on fast algorithm and LBP algorithm. Data Acquisit. Process. **32**(06), 1223–1231 (2017). https://doi.org/10.16337/j.1004-9037.2017.06.018 https://doi.org/10.16337/j.1004-9037.2017.06.018
18. Chen, X., Yu, F., Chen, Y.: Application of improved MB-LBP characteristics and LPP algorithm in strip surface defect identification. Sens. Microsyst. **39**(04), 156–160 (2020)
19. Huang, X., Ye, J., Xiong, J.: Road garbage image recognition and extraction based on texture feature fusion. Comput. Eng. Design **40**(11), 3212–3218, 3305 (2019). https://doi.org/10.16208/j.issn1000-7024.2019.11.025
20. Ling, X., Guo, R., Liu, G., et al.: Pedestrian detection with multiple feature cascade based on PCA. Manuf. Autom. **43**(03), 32–34, 76 (2021)

21. Ge, Y., Ma, L., Chu, Y.: Remote sensing image retrieval combined with discriminant correlation analysis and feature fusion. Chin. J. Image Graph. **25**(12), 2665–2676 (2020)
22. Li, M.Y., Ou, F.L., Yang, W.Y.: Correlation filtering video tracking with hog features. Data Acquisit. Process. **35**(03), 516–525 (2020). https://doi.org/10.16337/j.1004-9037.2020.03.014
23. Lin, J.Y., Huang, J.Q., Jiang, K.Y.: Phase error compensation for regional gamma pre-coding correction. Optoelectron. Eng. **43**(09), 32–38 (2016)

Decision Tree Fusion and Improved Fundus Image Classification Algorithm

Xiaofang Wang[1](✉) ⓘ, Yanhua Qiu[2], Xin Chen[2], Jialing Wu[2],
Qianying Zou[1] ⓘ, and Nan Mu[3]

[1] Geely University of China, Chengdu 641423, Sichuan, China
`939549393@qq.com`
[2] Chengdu College of University of Electronic Science and Technology of China,
Chengdu 611731, Sichuan, China
[3] Sichuan Normal University, Chengdu 610066, Sichuan, China

Abstract. In order to improve the effect of glaucoma fundus image classification, a new algorithm based on decision tree and UNet++ was proposed. Firstly, the image is divided into three channels of RGB, and the extracted green channel image is enhanced with the Butterworth parameter function of the fusion power function. Then the improved UNet++ network model is used to extract the texture features of the fundus image, and the residual module is used to enhance the texture features. The results of the experiment show that the average accuracy, the average specificity and the average sensitivity of the improved algorithm increase by 9.2%, 6.4% and 6.5% respectively. The improved algorithm is effective in glaucoma fundus image classification.

Keywords: Image classification · Butterworth parameter function · Improved UNet++ · Residual attention mechanism · Decision tree

1 Introduction

Glaucoma, as the second most common eye disorders in the world, is a group of eye conditions that damage the optic nerve [1]. It is the focus of fundus image separation detection research [2], and has attracted experts' attention.

Among them, He Xiaoyun et al. [3] put forward an improved UNet network model, which integrates residual block, cascade cavity convolution and embedded attention mechanism into UNet model to segment retinal vessels. Sabri Deari et al. [4] proposed a model of retinal vascular segmentation network based on migration learning strategy. The model was processed by pixel transformation and reflection transformation to enhance the data set. After processing, the retinal features were trained using the U-Net model to achieve retinal vascular segmentation; Yuan Zhou et al. [5] put forward the model of fusion attention mechanism and UNet++ network, based on UNet++ model, to achieve image feature extraction, and at the same time the attention mechanism is integrated

C. Yu et al. (Eds.): GPC 2022, LNCS 13744, pp. 35–49, 2023.
https://doi.org/10.1007/978-3-031-26118-3_3

into convolution unit to achieve feature enhancement, and then complete end-to-end detection of image; Ali Serener [6] and others proposed an image classification algorithm based on a single CNN convolution neural network model. This method can realize glaucoma image classification by creating multiple fusion of CNN. Guo Pan et al. [7] proposed a glaucoma image detection method based on MobileNet v2 and VGG classification network. The method used MobileNet v2 segmentation model to segment and locate the VGG image. Gupta et al. [8] proposed a method for detecting retinal vessels in random forest classification. The method could segment retinal images and extract the texture and gray features of the blocks, and then classify the retinal images. Ke Shiyuan et al. [9] used Support Vector Machine and Logistic Regression Integrated Multi-View Learning to predict glaucoma; DAS et al. [10] proposed a method for glaucoma detection based on CDR and ISNT rules. The method used region-growing method and watershed transform to segment OC and OD, and then realized glaucoma image classification. Narmatha Venugopal et al. [11] put forward a method of glaucoma image classification based on PH-CNN model, which uses DWT and PCA fusion method to extract features, and then uses PH-CNN model to classify glaucoma image automatically.

Although these algorithms can screen and judge glaucoma fundus lesions, but the accuracy of glaucoma fundus lesions detection is low, and the classification is not good. Considering these problems, this paper proposes an improved UNet++ algorithm for glaucoma image classification and detection. Firstly, the Butterworth parameter function is used to enhance the color, texture and contrast of the fundus image, and then the UNet++ network based on residual thought and mixed field of attention is used to extract the feature.

2 Algorithmic Implementation

In order to solve the problem of low contrast in glaucoma image classification, an improved UNet++ algorithm based on decision tree fusion is proposed. The whole algorithm is divided into three stages: preprocessing, feature extraction and image classification. The architecture is shown in Fig. 1.

Figure 1 shows that in the preprocessing stage, the green component image is extracted, and then the texture information and contrast of the glaucoma fundus image are enhanced by power function Butterworth parameter function; In the feature extraction stage, UNet++ network based on residual error module and attention mechanism is used; Decision tree C4.5 is used to classify the fundus images, and the results of glaucoma detection are obtained.

2.1 Preprocessing

Aiming at the problems of low contrast and poor detail information of the image, the improved Butterworth parameter function with power function fusion is used to preprocess the fundus image. The processing process is to first separate the RGB image, extract the green component image (as shown in Fig. 2), and then

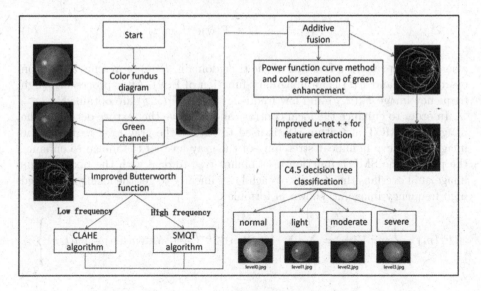

Fig. 1. Overall algorithm architecture diagram

use the improved Butterworth parameter function [12] to carry out frequency division processing to obtain high frequency information and low frequency information. The calculation is as shown in formula 1 and formula 2.

$$P_h = (R_h - R_l) \left/ \left(1 + \frac{aD_0}{D(x,y)}\right) \right. 2n + R_l \tag{1}$$

$$P_l = 1 - \left[(R_h - R_l) \left/ \left(1 + \frac{aD_0}{D(x,y)}\right) \right. 2n + R_l\right] \tag{2}$$

where R_h and R_l denote the high-frequency gain coefficient of the glaucoma fundus image and the low-frequency gain coefficient of the glaucoma fundus image. While $R_h > 1$, Represents high-frequency information that enhances fundus images.While $R_l < 1$, it is denotes reduced ocular fundus low-frequency information. a indicates the sharpening coefficient, D_0 indicates the cut-off frequency, n indicates the order of the filter, $D(x,y)$ indicates the distance between the frequency (x,y) and the filtering (x_0, y_0) center, and the Euclidean distance formula is adopted for calculation, as shown in formula 3.

$$D(x,y) = \sqrt[2]{(x - x_0)^2 + (y - y_0)^2} \tag{3}$$

After frequency division processing, high and low frequency information and low frequency information are obtained. In order to transform the information of high and low frequency into the image of high and low frequency for enhancement processing, inverse Fourier transform is used to transform the information of frequency domain into the image of spatial domain, as shown in formula 4.

$$F(t) = \frac{1}{2\pi} \int_{-\infty}^{+\infty} F(\mathrm{w}) e^{i\omega t} dw \tag{4}$$

where $F(t)$ represents a function of the time domain, $F(w)$ represents a function of frequency, and $F(t)$ is an imaginary function of $F(\mathrm{w})$. After processing, high frequency image $F_h(x, y)$ and low frequency image $F_l(x, y)$ are obtained.

In order to enhance the local contrast and improve the texture details of the image, the SMQT algorithm [12] is used to extend the gray level region of the image to achieve nonlinear stretching of the gray level of the image to enhance the pixels. The SMQT function uses binary tree to deal with the pixels of the image, and overlaps the outputs of each layer linearly to get the locally enhanced high frequency image, as shown in formula 5.

$$F_h'(m) = \left\{ m \mid V(m) = \sum_{l=1}^{L} \sum_{n=1}^{2^{l-1}} V(u(l,n)) 2^{L-l}, \forall \in M, \forall u(l,n) \in \forall u(l,n) \right\} \tag{5}$$

where m represents a pixel in image $D(m)$, $F_h'(m)$ is the output of SMQT, $v(m)$ is the gray value of the pixel, $U(m)$ is the quantization of the gray value, L represents number of layers in the binary tree, and n is the MQN output number with the number of layers of l.

In order to reduce the influence of color component on image detection, the low-frequency image is converted into Lab space [13], and the L-channel is processed by the method of histogram equalization. The processing principle is to divide the image into several blocks, classify each block, and interpolate each pixel with histogram equalization method to get the F_l' of the processed gray image. The processed high-frequency image and low-frequency image are fused to get the enhanced fundus image. The processing is shown in formula 6.

$$G(x, y) = aF_h'(x, y) + bF_l'(x, y) \tag{6}$$

where a and b denote the weighted constant respectively, and $G(x, y)$ represent the enhanced eye ground green component.

In order to reduce the influence of external factors and ensure the integrity of fundus image structure, the power function curve method [14] is used to reduce noise. The power function adjusts the image contrast mode by parameters and adjusts by image mapping relation. The calculation is shown in formula 7.

$$G' = ax^t + bx^{(t-1)} + \ldots \ldots + cx + d \tag{7}$$

where t is a power, it is a controllable parameter after the processing of enhanced image preprocessing.

2.2　Feature Extraction

UNet++ Networkm

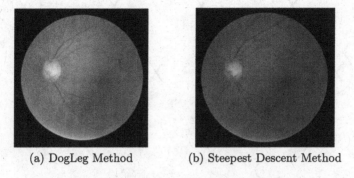

(a) DogLeg Method　　　　(b) Steepest Descent Method

Fig. 2. Comparison result diagram

In order to extract texture and detail information of fundus image effectively and improve the effect of image feature extraction, an improved UNet++ network model is proposed. In order to avoid the overfitting problem in the training process, the residual error idea is introduced into the model based on the UNet++ network. In addition, in order to improve the performance of the network feature extraction, the mixed domain attention mechanism is added to the residual module to enhance the texture feature. Its network architecture is shown in Fig. 3. As can be seen from the diagram above, the UNet++ network [15] consists of an encoder and a decoder, $x^{(i,j)}$ represents the output of the node $x^{(i,j)}$, where i represents the number of layers, j represents the j convolution layer of the current layer, black represents the original UNet, green and blue represent the dense convolution blocks on the jump path, and red represents depth oversight. The hopping path is used to change the connectivity of the encoder and decoder subnetworks. In UNet, the decoder receives the encoder's feature map directly, whereas in UNet++ there is a dense convolution block, and all convolution layers on the jump path use cores of size 3 by 3. Deep supervision enables the UNet++ model to run in both accurate and fast mode, averaging the output of all split branches. Quick mode is used to split the graph.Finally, only one branch is selected, and the results are used to determine the model pruning degree and speed gain.

Residual Module and Attention Mechanism

To solve the problem of gradient disappearance, the residual module [16] is introduced between the sampling and down sampling convolution layers on the UNet++ network, and a mixed domain attention mechanism [17] is added before each residual convolution module to obtain more local and global texture information. The architecture of the residual module is improved as shown in Fig. 4.

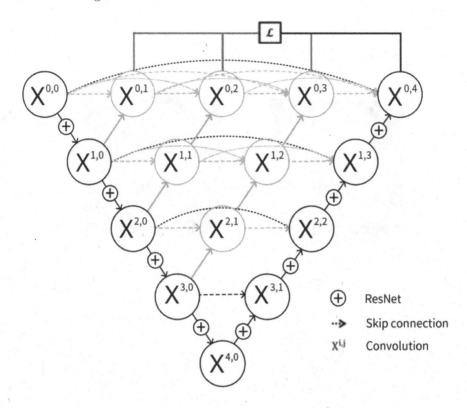

Fig. 3. Improved UNet++ model diagram

The principle of the residual module is to add the input feature graph and the feature extraction module to get the feature information, so that the network can contain the feature information of the input feature graph when propagating forward, and effectively solve the degradation problem of network model onvolution processing. The calculation is shown in Eq. 8.

$$H(x) = F(x) + x \tag{8}$$

where x represents the input network, $F(x)$ the feature extraction module, and $H(x)$ the output of the feature extraction of the eyeground image. In order to obtain specific feature texture information, the mixed domain attention mechanism is introduced into the residual error network. The network model in Fig. 5.

From Fig. 5, the input fundus feature map represents fed into the channel attention mechanism to perceive the global texture information, and the extracted information is fused with the original image to obtain the global feature processing results, as shown in formula 9.

$$CBAM\,(F_i) = SAM\,(CAM\,(F_i)) \times F_i \times (CAM\,(F_i) \times F_i) \tag{9}$$

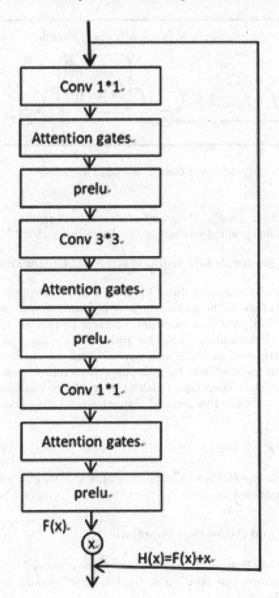

Fig. 4. Improved residual module structure diagram

where $CBAM(F_i)$ denotes the result of mixed domain attention mechanism, F_i denotes input fundus graph, $CAM(F_i)$ denote channel attention mechanism, SAM denote spatial attention mechanism, and × denotes matrix convolution.

The maximum pooling and the average pooling are then forwarded to a shared hidden layer MLP network. The dimensions of the two channel pooling maps are set to C × 1 × 1, and the result of the average pooling is processed by

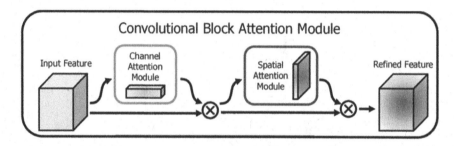

Fig. 5. Mixed domain attention mechanism

the sigmoid function. Finally, the result of the channel pooling is added to the result of the channel pooling mechanism as shown in formula 10.

$$CAM\,(F_i) = sigmod\,(MLP\,(AvgPool\,(F_i)) + MLP\,(MaxPool\,(F_i))) \quad (10)$$

where sigmod is the activation function, the $AvgPool$ is the average pooling process, the $MaxPool$ is the maximum pooling process, and the MLP is the MLP neural network, that is, the multilayer perceptual process, with the number of neurons in the hidden layer set to C/r and the r as a hyperparameter.

The spatial attention mechanism is pooled along the channel axis by average and maximum pooling, and then the two characteristic graphs are spliced together for convolution operation. Finally, the processing result of spatial attention mechanism is obtained by sigmoid activation, and the calculation is shown in formula 10.

$$SAM\,(CAM\,(F_i)) = sigmod\,(conv\,([AvgPool\,(\mathrm{M_c}) + MaxPool\,(\mathrm{M_c})])) \quad (11)$$

where SAM is the operation of spatial attention mechanism, and convolution operation is represented by $conv$.

Loss Function and Activation Function

To measure the classification error of the model, the improved UNet++ network uses the cross-entropy cost function as the loss function, and its calculation is shown in Formula 12.

$$C = -\frac{1}{n}\sum_x [y\ln a + (1 - y)\ln(1 - a)] \quad (12)$$

where y is the expected output of the glaucoma model, a represents the actual output of texture information, x represents sample data, and n represents the total sample size.

In order to optimize the network parameters, the stochastic gradient descent function is used as the optimization function. The principle is to select a part of dataset from the training dataset by random processing. Keep updating, get the final result, its parameter updating formula is shown in formula 13.

$$\theta_j = \theta_j \vartheta * \frac{\partial JS(\theta_j)}{\partial \theta_j} \qquad (13)$$

where S is the total number of training samples, and j represents the selected value of the samples, with a value of $\in [0, S]$.

The residual module is processed using PReLU [18] activation functions to reduce dependency on manual adjustments and reduce empiricism, as shown in Formula 14.

$$f(y_i) = \begin{cases} y_i, & \text{if } y_i \geq 0 \\ a_i y_i, & \text{if } y_i < 0 \end{cases} \qquad (14)$$

where i is different channels, for residual processing, each channel has different parameters PReLU function. a_i represents the initialization value, and 0.25 is the best.

The sigmoid function in the attention mechanism is an S-shaped saturation function, calculated as Formula 15.

$$sigmod(y_i) = \frac{1}{1 + \exp(-y_i)} \qquad (15)$$

2.3   Fundus Image Classification

To improve the classification effect of glaucoma fundus image and reduce the model training and detection practice, the multi-classification of glaucoma fundus lesions using c4.5 decision tree was studied [19].

The Decision Tree C4.5 algorithm is used to find split attributes from all texture information extracted from features, generate texture information and no texture information, continuously segment the nodes with texture information, and then classify glaucoma fundus lesions into normal, mild, moderate and severe glaucoma.

Decision tree C4.5 algorithm implementation is divided into two stages: initial decision tree generation and decision tree pruning. The algorithm flow is as follows:

input: Training set decision table: training set $DK = (d_1, k_1), (d_2, k_2), ...,$ (d_n, k_n) and set of attributes $AK = a_1, a_2, a_3, a_4$
output: Decision tree with Node as root node

1. function Build_DT(D,A) constructive function
2. generate node B
3. if samples is DK belong to the same category CK then
4. mark node as class CK leaf node; return
5. end if
6. if $AK = \emptyset$, samples in DK have the same value on AK then
7. mark B as a leaf node of the class with the most samples in DK return

8. end if
9. select the best attribute from AK, be$a_* = \arg\max_{a \in A} q_A^x GR(DK, a)$, attributes with the highest rate of gain
10. for a_* to every attribute value of a_*^v do
11. generate a branch for Node; make DK^v a subset of samples in DK that have a_* value of a_*^v
12. if DK^v is Null then
13. mark the branch node as the leaf node of the class with the most samples in DK
14. else
15. use Build_DT$(DK^v, AK\ a_*)$ as a branch node
16. end if
17. end for
18. end function

After the decision tree classification, the fundus image of glaucoma was detected to be normal, mild, moderate or severe.

3   Experimental Analysis

3.1   Experimental Environment

Data sets provided by Paddle Paddle were used and 480 glaucoma datasets were selected for training. Normal glaucoma, mild glaucoma, moderate glaucoma and severe glaucoma accounted for 120 each, as shown in Fig. 6.

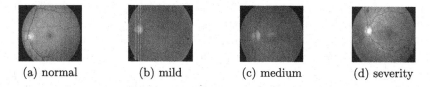

(a) normal   (b) mild   (c) medium   (d) severity

Fig. 6. Sample data set diagram

The study used Inteli7-7800 CPU, NVIDIA Ge Force GTX 1080i graphics card, Paddle Paddle 2G GPU, deep learning frameworks Keras, OpenCV and Tensorflow. Because the UNet++ network input layer requires 1024×1024 pixels, the Python pillow library is used to manipulate the crop to set a fixed clipping region to crop all images to 1024×1024 and train at a 7:3 scale.

3.2   Experimental Design and Analysis

Accuracy Acc, specificity S_p and sensitivity S_n were used to evaluate the classification of glaucoma fundus lesions, as shown in Formula 16, 17 and 18.

$$Acc = \frac{M + N}{M + N + L + P} \tag{16}$$

$$S_p = \frac{N}{N + P} \tag{17}$$

$$S_n = \frac{P}{P + L} \tag{18}$$

where M is the number of normal fundus maps, N is the number of glaucoma maps, L is the number of normal fundus maps, P is the number of glaucoma maps, and N and P represent the total number of glaucoma maps.

To obtain the global optimal effect of the loss function, the optimal learning rate of 0.001 is selected by adjusting the network weight super-parameters. The accuracy of different iterative experiments is analyzed during the model training, and the results in Fig. 7.

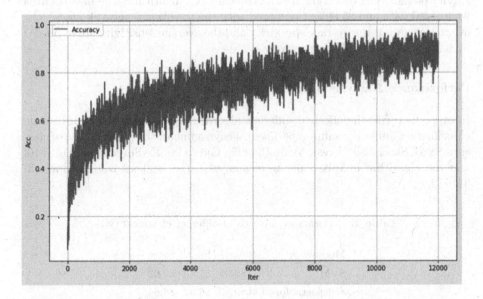

Fig. 7. Result chart of model average accuracy under different iterations

The results show that the average classification accuracy is the best when the learning rate is 0.001, and the average classification accuracy is 94.46%.

Analysis of Different Network Performance

In order to verify the effect of different algorithms on glaucoma fundus image classification under the same experimental environment, the accuracy, specificity and sensitivity of CNN, He [3], Sabri [4], Ali [6] and the algorithm presented in this paper are analyzed. The results are shown in Table 1.

Table 1. Performance analysis of different network models

Network	Acc	Sp	Sn
Classical CNN	78.86	79.12	81.32
He	88.03	89.25	93.78
Sabri	88.85	87.87	93.01
Ali	90.26	88.52	92.15
Proposed method	94.46	91.74	95.89

In Table 1, the average accuracy, the average specificity and the average sensitivity of glaucoma detection are all classical CNN algorithm. The best result is 94.46%, 91.74% and 95.89%. Compared with the traditional network model, the average accuracy, the average specificity and the average sensitivity are improved by 9.2%, 6.4% and 6.5% respectively.

Performance Analysis of Different Classifiers

In order to verify the effect of different algorithms on glaucoma fundus image classification under the same experimental environment, the performance of classical SVM, Stochastic Forest, Yuan Zhou [5], Gupta [8], Ke Shiyuan [9], DAS [10] and the algorithm in this paper were analyzed, the analysis results shown in Table 2.

Table 2. Performance analysis of different classifiers (%)

Method	Acc	Sp	Sn
SVM	89.92	87.85	89.02
Stochastic forest	89.80	86.32	88.64
YuanZhou	92.84	91.12	94.21
Gupta	90.76	83.30	96.10
Ke Shiyuan	92.24	89.13	95.50
DAS	91.47	88.82	92.15
Proposed method	94.46	91.74	95.89

It can be seen from Table 2 that the best is the algorithm studied in this paper and its accuracy, specificity and sensitivity of the improved algorithm are 94.46%, 91.74% and 95.89% respectively, which is 3.6%, 4.5% and 3.5% higher than the traditional algorithm. The improved algorithm has some advantages in fundus image detection.

Ablation Experiment

To verify that the proposed algorithm has a good effect on the detection of glaucoma lesions, a study was made on the ablation of the proposed algorithm under the same experimental environment, and the average detection results were evaluated using the accuracy rate, The results of the evaluation are shown in Table 3.

Table 3. Performance analysis of different classifiers (%)

Unet++	Butterworth filter	Power Function Method of Fusing Butterworth	Attention mechanism	Residual convolution module	Acc
					85.28
✓	✓				86.00
✓	✓		✓		86.59
✓	✓		✓	✓	87.21
✓		✓			87.88
✓		✓	✓		88.79
✓	✓	✓	✓	✓	94.46

This can be seen from Table 3, the traditional UNet++ network model is optimized by using the Butterworth power function and residual theory, which has some advantages in classifying glaucoma fundus lesions.

4 Conclusion

An improved UNet++ network model of C4.5 decision tree is proposed. The model can preprocess the green channel image, then enhance the texture of the feature image by using the improved UNet++ network. Finally, the decision tree was used to get the severity of glaucoma. The algorithm can be applied to classify and detect glaucoma fundus diseases, as well as other medical images and traffic images, but the training time and classification accuracy of the network model need to be further optimized.

References

1. Aich, G., Banerjee, G., Debnath, S., Sen, A.: Optical disc segmentation from color fundus image using contrast limited adaptive histogram equalization and morphological operations. In: 2021 International Conference on Smart Generation Computing, Communication and Networking (SMART GENCON), pp. 1–6 (2021). https://doi.org/10.1109/SMARTGENCON51891.2021.9645808
2. Li, X.: Research on image analysis and automatic disease diagnosis algorithm based on color fundus image. Beijing Institute of Technology (2017)
3. He, X., Xu, J., Chen, W.: Research on fundus blood vessel image segmentation based on improved U-Net network. J. Electron. Meas. Instr. **35**(10), 202–208 (2021). https://doi.org/10.13382/j.jemi.b2003781
4. Deari, S., Öksüz, İ., Ulukaya, S.: Importance of data augmentation and transfer learning on retinal vessel segmentation. In: 2021 29th Telecommunications Forum (TELFOR), pp. 1–4 (2021). https://doi.org/10.1109/TELFOR52709.2021.9653400
5. Yuan X, Guo H, Lu, L., Wei, L., Yu, Z.: High-resolution remote sensing image change detection algorithm based on UNet++ network and attention mechanism. J. Surv. Mapp. Sci. Technol. **38**(02), 155–159 (2021)
6. Serener, A., Serte, S.: Glaucoma classification via deep learning ensembles. In: 2021 International Conference on INnovations in Intelligent SysTems and Applications (INISTA), pp. 1–5 (2021). https://doi.org/10.1109/INISTA52262.2021.9548439
7. Guo, C., Li, W., Zhao, X., Zou, B.: Glaucoma screening method guided by semantic feature map. J. Comput. Aided Design Graph. **33**(03), 363–375 (2021)
8. Gupta, G., Kulasekaran, S., Ram, K., et al.: Local characterization of neovascularization and identification of proliferative diabetic retinopathy in retinal fundus images. Comput. Med. Imaging Graph. **55**(SC), 124–132 (2017)
9. Ke, S., Hu, M., Xu, Y.: Research on computer-aided glaucoma diagnosis algorithm based on ensemble learning. J. Beijing Univ. Chem. Technol. (Nat. Sci. Edn.) **46**(04), 86–91 (2019). https://doi.org/10.13543/j.bhxbzr.2019.04.013 https://doi.org/10.13543/j.bhxbzr.2019.04.013 https://doi.org/10.13543/j.bhxbzr.2019.04.013 https://doi.org/10.13543/j.bhxbzr.2019.04.013
10. Das, P., Nirmala, S.R., Medhi, J.P.: Diagnosis of glaucoma using CDR and NRR area in retina images. Netw. Model. Anal. Health Inform. Bio Inform. **5**(1), 1–14 (2016)
11. Venugopal, N., Mari, K., Manikandan, G., Sekar, K.R.: Phase quantized polar transformative with cellular automaton for early glaucoma detection. Ain Shams Eng. J. **12**(4), 4145 (2021)
12. Zhu, Z., Li, H., Zhao, L., Wang, H., Tan, J.: Research on image enhancement algorithm of magnetic tile surface defects based on improved homomorphic filtering and SMQT. Optoelectron. Laser **32**(08), 818–825 (2021). https://doi.org/10.16136/J.Joel.2021.08
13. Hao, M.: Research on underwater image enhancement method based on pyramid fusion. Dalian Maritime University (2020). https://doi.org/10.26989/D.CNKI.GDLHU.2020.000020
14. Wang, W., Chen, W., Wang, L., Yuan, J.: Research on a method of adjusting the contrast of fundus images. China Med. Equip. **12**, 11–13 (2006)
15. Zhou, Z., Rahman Siddiquee, M.M., Tajbakhsh, N., Liang, J.: UNet++: a nested U-net architecture for medical image segmentation. In: Stoyanov, D., et al. (eds.) DLMIA/ML-CDS -2018. LNCS, vol. 11045, pp. 3–11. Springer, Cham (2018). https://doi.org/10.1007/978-3-030-00889-5_1

16. Fu, Y., Li, Z., Zhang, Y., Pan, D.: A U-net with residual module for detecting the exudation of fundus images. Minicomput. Syst. **42**(07), 1479–1484 (2021)
17. Chen, H., Zhen, X., Zhao, T.: A small target detection model based on channel-spatial attention mechanism feature fusion [J/OL]. J. Huazhong Univ. Sci. Technol. (Nat. Sci. Edn.) 1–8 (2022). https://doi.org/10.13245/J.HUST.238491
18. Chen, X., Well, L., Zhang, Y., Xue, Q.: Post-nonlinear BSS algorithm based on multilayer neural network and PReLU function. J. Mianyang Norm. Univ. **39**(02), 25–30+46 (2020). https://doi.org/10.16276/j.cnki.cn51-1670/g.2020.02.006
19. Fang, C., Fan, M.: Text detection based on improved convolutional neural network and line features. Microelectron. Comput. **36**(08), 77–82 (2019). https://doi.org/10.19304/j.cnki.issn1000-7180.2019.08.017

Construction Method of National Food Safety Standard Ontology

Die Hu, Chunyi Weng, Ruoqi Wang, Xueyi Song, and Li Qin[✉]

College of Informations, Huazhong Agricultural University, Wuhan 430070, China
qinli@mail.hzau.edu.cn

Abstract. The food safety standard ontology is the schema layer of top-bottom building method for constructing the food safety knowledge graph, which can effectively ensure the professionalism and effectiveness of the knowledge graph. However, there are few relevant research in recent literature, which also restricts the application of knowledge graph in the food safety field. In order to solve this problem, this paper proposes the construction method of food safety standard ontology based on the national food safety standards. Firstly, we built the class framework of the food safety standards according to the seven steps of ontology construction, and then entities, attributes and relations between entities are extracted from the national food safety standards by employing the rule-based knowledge extraction algorithm. Finally, we import those entities, attributes and relations into the class framework to complete the ontology. According to the experimental results of our entity mapping, the food safety standard ontology can describe the mainly important concepts, terms, operation process and their relation in the standards. So it will effectively support the construction and integration of knowledge graph.

Keywords: Food safety standards · Domain ontology · Class framework · Knowledge extraction algorithm

1 Introduction

In recent years, food safety issues often attract people's attention. As food safety issues are closely related to the life of every person, frequent outbreaks of food safety accidents will affect the people's expectations for the future economic and social development, and it will affect the long-term stability of the society [1]. Food safety problems mainly include: pathogenic microorganism pollution, pesticide and veterinary drug residues, heavy metal and mycotoxin pollution, illegal and adulterated use of food additives [2]. To solve these problems, many countries formulated relevant standards to regulate the food from food raw materials, processing, packaging, transportation and sales processes. However, after years of revision and supplement, the content of food safety standards is

This document is supported by the China Fundamental Research Funds for the Central Universities (No: 2662022XXYJ001, 2662022JC004).

messy and complex, if regulators and producers do not fully understand food safety standards, they will "ignorant" use some harmful technology and production environment for production, and make it into the market, causing harm to consumers [3]. For consumers, the scale of Internet data has grown exponentially and disorderly, which makes the acquisition and utilization of food safety knowledge become more difficult, all these create food safety information asymmetry between the government and consumers. Therefore, it is necessary to construct a large and standardized food safety ontology for producers and consumers to use. Moreover, it can also provide a schema basis for the top-down construction of food safety knowledge graph.

Using ontology to describe data semantically can not only excavate the essential meaning of data, but also improve the retrieval efficiency [4]. Moreover, ontology provides the pattern structure of the top layer of knowledge graph, which can greatly assist the construction process of knowledge graph [5]. We have done some research on the food safety knowledge graph [6–8], but the food safety ontology is still insufficiently. In those studies the bottom-up method is used to build the knowledge graph, that means, entities, attributes and relations in documents are extracted only through machine learning algorithm and entity merging only uses conceptual similarity. The top-down knowledge graph construction refers to the per-construction of ontology, which can improve the quality of knowledge extraction and facilitate to the integration of knowledge graph in different fields. In view of this, this paper attempts to explore ontology from national food safety standards of China. The establishment process is to mine the class framework and class relations in the standards, and then extract the instances and attributes of classes by our knowledge extraction algorithm, finally verify the validity of ontology by entity mapping.

2 Related Research

For computer science and information science, ontology is a formal, distinct and detailed description of shared concept system. According to the widely accepted definition of ontology [9], ontology is a normative and unified definition of conceptual forms and relations between concepts.

The biggest feature of ontology is sharing, and the knowledge reflected in ontology is a well-defined consensus [5], then ontology is a knowledge concept template used to describe the concept hierarchy. Ontology modeling is a very important phase in the process of knowledge graph construction, which can well sort out entities and relations among entities. In the process of building knowledge graph, the introduction of ontology will play a guiding role, which can help the machine to understand the relation between concepts and make knowledge reasoning smoothly. It is worth noting that, when studying the domain ontology model, we should follow the principles of completeness and consistency put forward by Gruber, and construct the domain ontology reasonably and normatively [10]. In addition, the research on the construction of domain ontology can be divided into two levels according to whether reusing the existing mature metadata set, the first level is reference to mature ontology, the second level is building a new domain ontology.

In food safety field, there has few research on ontology, Dooley established a comprehensive and easy-to-access domain ontology of food from farm to table (FoodOn), which accurately and consistently described the common foods in cultures around the world [11]. Li Hongwei analyzed the early warning information of food safety, and built a ontology of Hazard Analysis Critical Control Point (HACCP) [12]. The food safety standards are related to the food production process, and currently has no normalized dataset can be used. Therefore, we collected the current national food safety standards of China as data source, and analyzed its content and extracted knowledge to build ontology.

3   Model Design of Food Safety Standard Ontology

The construction process of our domain ontology mainly includes several part, that are model design of ontology, natural language process, knowledge extraction of entities and knowledge import. The specific process is shown in Fig. 1.

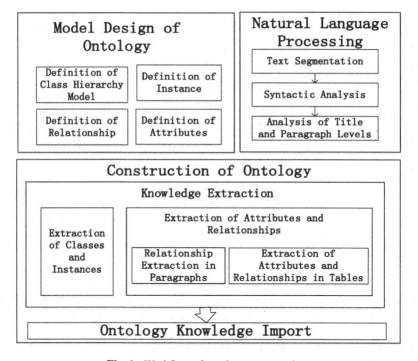

Fig. 1. Workflow of ontology construction.

3.1   National Food Safety Standards

Food safety standards are technical requirements and measures formulated after risk assessment of chemical, biological and physical substances that exist or may exist in food, food-related products and food additives. It is the most basic requirement for food to enter the market, and it is the technical regulations that should be implemented in accordance with food production and operation, inspection, import and export, supervision and management, and an important basis for food safety supervision and management [13, 14]. In our ontology study, the national food safety standard of China is the the only food safety standard enforced by the state.

We obtained 1182 documents about national food safety standards from FoodMate (http://www.foodmate.net). In our research the current national food safety standards are classified according to their content and scope of application. He Xiang [15] divides the national food safety standards into general standards, production, operation standards, inspection methods and procedures and food-related product standard. The national health administrative department has also established and formed the national food safety standard system [16] which includes four categories: general standard, product standard, product operation standard and detection method standard. Here, the standards in this paper are classified by the official classification scheme.

3.2 Model Framework of Food Safety Standard Ontology

The methods of ontology modeling are different according to different application purposes. At present, the mainly methods are the following two ways: the first is to explore the construction method from the perspective of knowledge engineering, which can be called ontology engineering; the other is the semi-automatic construction method of transforming the existing thesaurus resources into ontology. In addition, Ding Shengchun [17] proposed a comprehensive (semi-automatic) ontology construction method based on top-level ontology. The mainstream construction methods of knowledge engineering methods include Seven Steps method [18, 19], Skeleton method [20, 21], Methodology method [22], KACTUS engineering method [23], SENSUS method, IDEF-5 method. In this paper, Seven Steps method was used, Seven Step method is proposed by Stanford University School of medicine, which is mainly used to construct domain ontology.

Class Hierarchy Model of Food Safety Standard Ontology
In our research, the top-down construction mode [24] is adopted to build the ontology class model, that is, the top layer class (parent class) is abstracted first, and then their subclass are found out. If the subdivision can be continued, the third layer class will be established and refined step by step. Some class model are shown in Fig. 2.

Top-Level Class Model
Food category entities, as the most important concept described in food safety, are frequently cited in all food safety general standards. For example "GB 2760-2014, Standard for The use of Food Additives" specifies the types and contents of additives allowed to be added in each food category. Based on its frequency of occurrence, we regard "food category entity" as the top-level class in food standard ontology. Besides, "name of food safety standard" as the label of each standards was also defined as another top-level class of food safety ontology. Then by analyzing the content of food safety standards, the phrases "terms and definitions" are frequently appeared in various standards. The content under "terms and definitions" is the meaning of some professional terms in food

field, so we defined "standard terminology" as the third top-level class. Finally, for further show the specific content of the standard, "content of food safety" was extracted as the fourth top-level class.

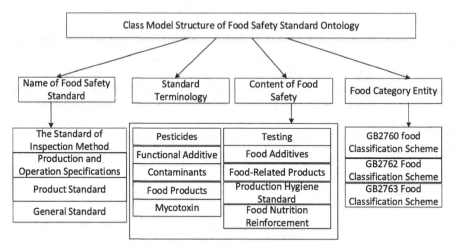

Fig. 2. Framework of top class and some subclasses

Secondary-Level Class Model

After completing the design of the top-level class, we continue analyzes its sub-class, that is, the second-level class. The model of the top-level class and its sub-class is shown in Fig. 2.

(1) Food category entities

In the existing GB/T national food safety standards, we have found three different food classification schemes, which are respectively derived from GB 2760-2014 Standards for the Use of Food Additives, GB 2763-2019 Standards for Maximum Residues of Pesticides in Foods and GB 2762-2017 Standards for Pollutants in Foods. In order to distinguish the three different food classification systems, in this paper, these three sets of food classification schemes are set up as secondary-level class, that are "GB2760 food classification scheme", "GB2763 food classification scheme" and "GB2762 food classification scheme". They are all subclass of "food category entities". In addition, all the food categories defined in the three classification systems are instances under its subclass.

(2) Name of food safety standard

At the beginning of our study, all food safety standards has been classified into different types, including general standards, production and operation standards, inspection method standards and product standards. Because each category is quite different in standard content and format, the extracted instances and relations are also quite different. So the names of these categories are also regarded as subclass of "name of food safety standard".

(3) Content of food safety

The content of food safety frequently include several common inspection items, such as food additives, pesticides, etc., and the actual extraction process proves that most standards are defined around these topics, such as the quality specifications and standards of food additives, physical and chemical testing standards. This shows that use these keywords as subclass are reasonable and feasible.

Through the process of "(1) analyzing the standard content, (2) design class name, (3) tracing back to the content, (4) determining the class name", secondary-level class of all food safety contents of the standards are extracted.

Other Subordinate Classes of Ontology

Under the secondary-level class model, you can continue to define subclass from the detailed description of standard content. For example, according to the content of production hygiene standards, as the secondary-level class, production hygiene standards include the following third-level subclass: site selection and production environment, factory buildings and workshops, requirements of raw materials and packaging materials, facilities and equipment, product traceability and recall, product hygiene standards, management system and personnel, records and documents management, etc. Factory buildings and workshops can be further subdivided, such as general requirements of factory building and workshop, special factory buildings design requirements, design and layout, workshop temperature control, etc.

Composition of Food Safety Standard Ontology

After determining the class model of the ontology, it is also necessary to determine the attributes of the class and the member instances of the class so as to realize the construction of the ontology.

Instance in Ontology

In ontology, instances represent the realization of classes, for example, propylene glycol is an instance of food additive, and silicon dioxide is also an instance of food additive. According to the analysis of the content of food safety standards, the instance definition rule are designed as follows:

(1) For the class of "food category", all names of food category in the "GB 2760" are the member instances of subclass "GB 2760, food classification scheme", such as "01.0, milk and dairy products", "01.02, fermented milk and flavor fermented milk", etc. Similarly, instances of other classification schemes are also determined according to this idea.
(2) Most instances of food safety testing items appear in general standards, usually a document hold a kind of instances. For example, "GB 2760-2014, Standard for the Use of Food Additives" lists the varieties of food additives that are allowed to be used in foods, and "GB 2761-2017, limit of mycotoxins in Foods" lists mycotoxins that may pose a greater risk to public health. Pollutants that may pose greater risks to public health are listed in "GB 2762-2017, Limits of Pollutants in Food", and the specific items listed in each standard are instances of corresponding categories.

(3) The names of all China food safety standards are instances of the corresponding categories of the standards. For example, "GB 5009.7-2016 National Food Safety Standard Determination of Reducing Sugar" stipulates the determination method of reducing sugar, which belongs to the inspection method category, that is, "GB 5009.7-2016, National Food Safety Standard Determination of Reducing Sugar" is an instance of "inspection method standard". Also the set of terms and nouns listed below the heading of "Terms and Definitions" in each food safety standard is the set of member instances of "standard terms".

Relations in Ontology

(1) Relations between instances

Relations definition needs to analyze the actions of various standards. First, we defined the relations in every general standards, and then the relations in other standards would be defined. The specific definition method is described as follows:

① General standards usually involve instances of food categories and food sampling inspection items, such as "benzoic acid and its sodium salt" is the instance of class "food additives" and "Blended Soy Sauce" is the instance of "GB2760 food classification scheme". There is a relation <has_FoodAdditive> between "Blended Soy Sauce" and "benzoic acid and its sodium salt", in which <has_FoodAdditive> is the relation name; However, "legume vegetables and potatoes" which is the instance of "GB 2762 Food Classification scheme" may contain the pollutant "lead", this can be described as ("legume vegetables and potatoes" <has_Pollutant> "lead"), among which <has_Pollutant> is the relational name.

② Other standards, such as quality specifications of food additives and related standards, have ("instance of measurement items" <has reagents and materials> "instance of reagents and materials"), such as ("determination of total lactic acid" <has reagents and materials> "sulfuric acid"), ("determination of aluminum trioxide" <has instruments and equipment > "spectrophotometer"), the instances of the last relation are instances of "method principle" which are the subclass of "Testing" class.

③ There is also a reference relation between the standard name and the standard content specified by it. we defines a pair of reciprocal attributes of <prescribe> and <is _ prescribed _ by>. For example, "GB 10133–2014, National Food Safety Standard Aquatic Condiments" is a regulation on aquatic condiments, then the constraint ("GB10133–2014, National Food Safety Standard Aquatic Condiments" <prescribe> "Aquatic Condiments") would be added to the ontology. If the subject is exchanged, then use the relational name <is _ predicted _ by>. However, for the standard terms, the standard terms are quoted by the standards, so the relation between them is named <has_term> and <is_term_of>, such as ("GB12694-2016, national standard for food safety, hygienic standard for livestock and poultry slaughtering and processing" <has_term> "clean area"). For non-universal standards, the more common relation name is <has_content>, which are used to represent the specific project regulations in those standards. Some reference relations in GB 8950–2016 are show in Table 1.

(2) the relation between similar instances.

There can also be a relation between instances of the same class, and it is often a pair of reciprocal relations. For the instance of class "food category", we defines each food category name in all food classification scheme as an instance, but since they are categories, there should also be a kind of relation between them, and this relation is <parent class> and <subclass>. For example, in the food classification scheme of GB2760, "01.0 milk and dairy products" is a food category class, and "01.02 fermented milk and flavor fermented milk" is a subclass of "01.0 milk and dairy products". They are all instances of "food categories", and there is an inheritance relation between them. Then add the constraints ("milk and dairy products" <subclass> "fermented milk and flavor fermented milk") and ("fermented milk and flavor fermented milk" <parent class> "milk and dairy products") in the ontology.

This kind of relation also exists between standard term instances. In order to distinguish the relation between standard term instances and food category instances, we defines reciprocal relation names as <has_term> and <is _ term _ of>, such as ("pollen" <has _ term> "bee pollen").

Table 1. Reference relations in GB 8950–2016

Standard name	Relation	Instances
GB8950-2016	prescribe	Canned food
GB8950-2016	has_range	This standard applies to the production of canned food
GB8950-2016	has_term	Commercial sterilization
GB8950-2016	has_content	The structure of equipment, tools and fixtures used in the canned food processing workshop and the installation position of fixed equipment should be convenient for thorough cleaning and disinfection

Attribute in Ontology

Because attribute values can be integers, floating-point numbers, Boolean values and strings, in order to describe an instance more accurately, attributes are often used to represent the characteristics of an instance. For example, ("benomyl" <ADI> "0.1mg/kg bw"), where ADI is an attribute name and 0.1mg/kg bw is its corresponding value.

For additives such as food additives and pesticides, they also have their own characteristic attributes. For example, food additives have CNS number, INS number, sensory requirements and other attributes, while pesticides also have numeric type attributes such as pesticide residues and ADI value of daily allowable intake. The specific contents are shown in Table 2.

Table 2. Attributes of additives

Subject class	Subject	Attribute	Object
Food additives	Calcium hydrogen phosphate	CNS number	06.006
Food additives	Calcium hydrogen phosphate	INS number	341ii
Food additives	Calcium hydrogen phosphate	Color requirement	White
Pesticide	Benomyl	ADI	0.1 mg/kg bw

4 Data Import of Food Safety Standard Ontology

4.1 Data Extraction

The creation and maintenance of ontology often takes a lot of time. Youn Jason et al. proposed a semi-supervised framework of automatic ontology population from the existing ontology support by using the method of word embedding [25]. In our research, a rule-based optimal semantic matching algorithm is adopted to realize semi-automatic knowledge extraction.

The knowledge extraction rules we defined rely on keyword analysis. The key words are recurring in the food safety standards, such as maximum limit, instruments and equipment, reagents and materials, etc. we obtained keywords through statistical word frequency and semantic analysis. In general, the document format of the same type standard is approximately the same, and the content needs to be extracted also has similar core words. Therefore, in the process of automatic extraction, the same type standards will be extracted in batches, and obtain all the class, attribute and instances. The following two methods are used to mine the instance relation:

(1) For unstructured text, in order to identify the document structure, context relation should be kept as much as possible, and the relation between instances should be found through the context relation, those relation usually is <include> or <has_content>.
(2) For the structured data in the table, the relation between instances is discovered through the header of the table, and the relation name is usually the field name in the header.

Furthermore, the document layout is mainly judged by the subtitles and title numbers of all levels, and the title numbers can be considered as the relations between classes, such as the inclusion relation and the instance relation.

Extraction of Classes and Instances

In food safety standards, all levels of subtitles often contain core words. we puts these words into the class name pool as candidate class names. The text under the all subtitles is taken as an instance of title class. However, when the all levels of subtitles and the text behind the subtitles are actually extracted, there will be a problem that the layers of title number in different paragraph is not same. Because, there are probably have two-levels titles, three-level titles or more in the text, and sometimes there are no text behind the subtitle. Therefore, we must design algorithms to automatically identify the layouts of the document, so as to extract the title classes and content instances more accurately. The relation name between a title class and its subtitles class is <include>.

The algorithms of document layout identification use digital label to mark the all levels of subtitles and text behind them. For example, the marking rules of document layout of "GB 8955-2016, National Food Safety Standard, Hygienic Specification for the Production of Edible Vegetable Oils and Their Products" is listed as follows:

(1) If the title number is subtitle, for example, there is the subtitle "4.1 General Require-ments" under the title "4 Workshop and Workshop", and there is a <include> relation between them.

(2) If there is no subtitle after a title, for example, if the next paragraph behind subtitle "4.1 General Requirements" dose not have a subtitle number, then the text in the next paragraph is a description of the subtitle, and we considered them as an instance of the subtitle "4.1 General Requirements".

(3) Marking for each class and instance. For example, the title candidate class "4 factories and workshops" is the first-level title, marked as 1; The candidate title subclass "4.1 General Requirements" is a second-level title, marked with 2; The text behind "4.1 General Requirements" is its instance, and there is a <has_content> relation between "4.1 General Requirements" and its instance, so it is marked as 3.

(4) The title candidate subclass "4.2 Design and Layout" is the same level with the title candidate subclass "4.1 General Requirements", so it is marked as 2. In this way, each title class would be marked.

(5) If there is no subtitle and no text behind a title, for example, subtitle 5.1 and subtitle 5.2 has title numbers, but there is no text and no subtitle behind them, so they can be regarded as instances of the title candidate class "5 Facilities and Equipment".

After the above marking process, all classes and instances need to be extracted by marking values. In order to save the paths of all classes and instances, we try to use the stack as an intermediate medium to extract all data. The specific method are as follows:

(1) According to the marking values, the marked data in the document are stored in the stack in turn. Each data in the stack has an inclusive relation from the bottom of the stack to the top of the stack, so their marking values are also increasing in sequence. When the marking values are not increasing, the next data may temporarily suspend store, and the label of the top data of the stack would be recorded, which is an instance of the previous data. Then store all the data in the stack in an ordered list, which saves a path from vertex to leaf in the tree, this path also reflects the < include> relation from class to its subclass and its instance.

(2) When a new data re-enters the stack, it is necessary to check whether the marking value of new data is greater than that of the top data in the stack. If not, the data in the stack need to be popped out in turn until the marking value of the top data is less than the newly added data. Then the new data is pushed into the stack.

(3) return to step (1).

For example, title "4 workshops and workshops", subtitle "4.1 general requirements" and text after "should meet ......" are stored in the stack in sequence, with marking values 1, 2 and 3 respectively. At this time, the next data "4.2 design and layout" that needs to be pushed into the stack which has a marking value 2. When the data were judged needs to be popped out of the stack, the sequence of the data in the stack will be saved in a order list. At the same time, the relation between the the standard name class and its instance need to be retained, so the standard name class needs to be inserted into the order list in the head. Then the final sequence is ["GB 8955-2016, National Food Safety Standard, Hygienic Code for the Production of Edible Vegetable Oil and Its Products", "4 factories and workshops", "4.1 general requirements" and "should meet ......"]. After the path being saved, the data with marking values 2 and 3 would be popped out from stack and new data "4. 2 Design and Layout" would be pushed into the stack, and then repeat the previous operation with the marking value. The specific algorithm description is shown in Algorithm 1.

Algorithm1:   Paths extraction method of classes and instances

Input: Paragraphs in document C = $\{c_1, c_2, ..., c_n\}$, n is the total of paragraphs.

Output: Instances paths from upper classes.

1. Text layout marking:

Set the marking values as L = $\{l_1, l_2, ... l_n\}$

$i = 1$

For $i < n$ {

Read c_i,

   If c_i contains the P-level title number {

     $l_i = P$, $P \in (1,2,3..m)$, where m is the maximum number of all levels.

   } else if c_{i-1} contain a P-level title number {

       $l_i = l_{i-1} + 1$

    }else{ $l_i = l_{i-1}$ }

  $i = i+1$

}

2. Stack initialization: Create a stack and empty it, and set the pointer *cur* to point to the bottom of the stack;

3.   Save the path for all classes and their instances:

$i = 1$

For $i < n$ {

Reading c_i and l_i

If *cur* points to the bottom of the stack {

   push c_i and l_i into the stack

   cur points to c_i and l_i (current top of stack)

} else if $l_i > l_{cur}$ {

    push c_i and l_i into the stack

    cur points to c_i and l_i (current top of stack)

  } else{

    read the standard name of the document, and joint with the data from the bottom of the stack to the top of the stack in turn, and store path in a sequential list.

    While $l_i <= l_{cur}${ pop out the top element}

    push c_i and l_i into the stack

    cur points to c_i and l_i(current top of stack)

    }

  }

  $i = i +1$

save the path.

}

4. Get the instances: Search each path list, and the last element of the order list is the instance.

Extraction of Attribute and Relations

Relations Extraction from Documents

For we have defined <include> relations between classes, also between classes and instances, and algorithm 1 has saved the instance paths we need, so the <include> relations between them can be obtained by reading the paths. In order to distinguish whether the relations are between classes or between classes and instances, algorithm 2 stores the <include> relation in two tables, named as "class_ relations" and "classes_ instances" respectively.

Algorithm 2: Save < include > relation between classes and instances.

Input: Path for all classes and their instances

Output: "class_relations" between classes, and "class_instances" between classes and instances.

1: initialization: "class _ relations", "class _ instances" are all empty lists.

2. Get the relation between classes:

For each path {

   $i = 1$

   For $i <$ length(path)-1 {

      relation = (Path[i-1],subclass,Path[i])

      save the relation in the class_relations list

      $i = i+1$

   }

}

3. Get the relation between class and instance:

For each path {

   relation = (Path[-2],instance,Path[-1])

   save the relation in the class_instances list

}

Attribute Extracting from Tables

Tables appearing in standards usually have a table header, which defines the attribute meaning of the table. we matched the table header to the class name and instance name and judged whether the data in table have the relation we need. If the data in table is necessary, we extracted attribute from tables. In the table, the header is the semantics of each attribute, it is also the relation name of class and its instance. So if the relation triplet in table is (? S,? P,? O), predicate (? P) is the header name and the object (? O) is the attribute value, at last the content of subject (? S) needs to be analyzed through table data.

For example, we extracted relational triples in Table 3. The relational triples are ("calcium hydrogen phosphate "<color requirement> "white"), ("calcium hydrogen phosphate" <state requirement> "powder"), ("calcium hydrogen phosphate" <color inspection method> "take appropriate samples"), ("Calcium hydrogen phosphate" <State inspection method> "Take appropriate sample"). It can be found that the relation name in this table is actually the attribute value (such as color and state) joint with the table header (such as requirements). While the subject of the relation triple is the instance (such as calcium hydrogen phosphate) in the food safety standard. Therefore, the extraction method of relational triples in tables is as follows:

(1) First, read the instance in the food safety standard. we did it by segmenting the name of food safety standard, then matched keywords in the instance name set;

(2) Then jointed the relation names;

(3) Finally, found the corresponding relational attribute values. See algorithm 3 for specific implementation.

Table 3. Sensory requirements in "GB 1886.3-2016, National Food Safety Standard, Food Additive Calcium Hydrogen Phosphate"

Project	Ask	Test Method
Colour and lustre	White	Take a proper amount of sample and put it in a clean and dry white porcelain plate, and observe its color and state in natural light
Condition	Powder	Take a proper amount of sample and put it in a clean and dry white porcelain plate, and observe its color and state in natural light

---

Algorithm 3 Save the attributes relations in table

Input: Tables in standards
Output: Attribute and relation
1. Initialization: attribute relations is an empty list.
2. Get the instance *obj*: get the *obj* name from the instance name by word segmentation and string matching.
3. Get the attribute relation: extracting the values of attributes, row[i, j] means the value of row i and column j, and when $i=0$, it point to the header row .
$i = 1$
For $i < = R$ {//R is the maximum number of records in the table.
    relation = (obj,row[i,S]+row[0,V],row[i,V])
    //S is the column number where the prefix name of the relation is located, and V is the column number where the relation value is located.
    save relations
}

4.2 Data Import

Finally, the entities and relations are imported into the class model framework. When all entities and relations are imported into the ontology model, ontology visualization can be shown. Figure 3 show an diagram of part instances and relation. In the Fig. 3, the green inner ring is an instance of class "food safety standard name". Actually, they belong to different classes, but for the size limit of the diagram, the class is not shown here. And the gray outer circle instance is instance of the "range", in which the dotted line points from the instance of "standard name" to the instance of "range". Those dotted line indicated that there is a relation between them, and the name of the relation can be viewed by clicking the dotted line.

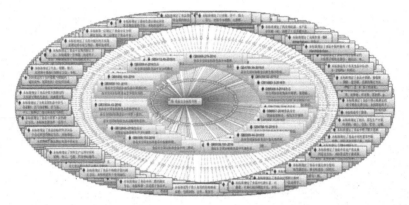

Fig. 3. Entity and relation diagram of food safety standard ontology (part)

4.3 Validity Analysis

In order to verify the correctness and effectiveness of the ontology we have built, we mainly use two methods: one is to invite research experts in food safety to manually modify the data and structures in the ontology; the other is to use the entities mapping method to do concept mapping with some food name corpora we have collected. Here, we use the NER technology to achieve this, and then we need obtained the relationship between the food name entities and the food category class or subclass in the ontology, Here we adopt C-norm [26] (a new share neural method to resolve the few shot learning entity linking problem) as our classifier to achieve this goal. At last we manually identify the mapping results to determine the precision and the recall of the entity mapping. My experimental results show that the precision and the recall is 0.85 and 0.72.

5 Summary

For a long time, food safety knowledge lack of standardization which will lead to ambiguity in understanding, though the study of food safety ontology is an effective way to solve this problem, but few researches were done on it. We collected 1182 national food safety standards from web, and build class framework for them. Then a new rule-based knowledge extraction algorithm was proposed to extract the instances and relations in all standards. The whole ontology includes 236 classes, 48 relation names, 823 attribute names, 8812 instances and 131,406 constraints (relations).

Our ontology describes various concepts and semantic relations related to food safety stipulated in national food safety standards, including food classification system, additive limits of various foods, pesticide and veterinary drug residues, pathogenic microorganism pollution, heavy metal and mycotoxin pollution, and food inspection, detection and physicochemical analysis methods. The significance of ontology is to provide schema layer for the top-down construction of food safety knowledge graph.

In order to verify the effectiveness of ontology, we use manual verification and entities mapping experiments to prove the effectiveness of our ontology. The current accuracy and recall can meet our basic needs, we will adopt BERT model to further improve them in future study.

References

1. Xu, X.: Food safety in China: problems, causes and countermeasures. Issues in Agric. Econ. **23**(10), 45–48 (2002)
2. Xu, R., Pang, G.: Study on the current situation, problems and countermeasures of food safety in China. China Science Publishing & media Ltd. (CSPM), BeiJin (2015)
3. Zhou, D., Yang, H.: Information asymmetry and government supervision mechanism in food quality and safety management. Chin. Rural Econ. **6**, 29–35, 52 (2002)
4. Gao, X.: Research on the construction and analysis method of domain Ontology of film and television works. Zhejiang Normal University of China (2014)
5. Liu, Q., Li, Y., Duan, H., Liu, Y., Qin, Z.: Knowledge graph construction techniques. J.Comput. Res. and Dev. **53**(3), 582–600 (2016)
6. Qin, L., Hao, Z., Li, G.: Construction and correlation analysis of national food safety standard graph. J. Comput. Appl. **41**(4), 1005–1011 (2021)
7. Li, Q., Zhigang, H., Liang, Z.: Food safety knowledge graph and question answering system. In: ICIT 2019: Proceedings of the 2019 7th International Conference on Information Technology: IoT and Smart City, pp.559–564. ACM, New York(2019)
8. Qin, L., Hao, Z.: National food safety standard graph and its correlation research. In: Liang, Q., Wang, W., Mu, J., Liu, X., Na, Z., Cai, X. (eds.) Artificial Intelligence in China. LNEE, vol. 653, pp. 405–411. Springer, Singapore (2021). https://doi.org/10.1007/978-981-15-8599-9_47
9. Gruber, T.R.: A translation approach to portable ontology specifications. Knowl. Acquis. **5**(2), 199–220 (1993)
10. Gruber, T.R.: Towards principles for the design of ontologies used for knowledge sharing. Int. J. Hum.-Comput. Stud. **43**(5–6), 907–928 (1995)
11. Dooley, D.M., Griffiths, E.J., Gosal, G.S., et al.: FoodOn: a harmonized food ontology to increase global food traceability, quality control and data integration. NPJ Sci. Food **2**(1), 1–10 (2018)
12. Li, H., Huang, W., Hong, X.: Research on ontology building in food security pre-warning. Comput. Technol. Dev. **23**(9), 238–240, 244 (2013)
13. Chen, J., Li, B.: China's food safety standard system: problems and solutions. Food Sci. **09**, 334–338 (2014)
14. Wang, H.-X., Jian, F., Zhang, L., Huang, Q.: Analysis and suggestion on the current situation of standards in food inspection in China. J. Food Saf. Qual. **011**(006), 2001–2006 (2020)
15. He, X.: Research on national food safety standard system construction. Central South University (2013)
16. Letter of the general office of the national health and Family Planning Commission on notifying the conclusion of the cleaning and integration of the catalogue of national food safety standards and food related standards. Chinese Journal of Food Hygiene (2017). CNKI:SUN:ZSPZ.0.2017-04-034
17. Ding, S., Li, Y., Gan, L.: Research on integrated domain ontology construction method based on top-level ontology. Inf. Stud. Theory Appl. **30**(2), 236–240 (2007)
18. Swartout, B., Patil, R., Knight, K., et al.: Toward distributed use of large-scale ontologies. In: Proceedings of the Tenth Workshop on Knowledge Acquisition for Knowledge-Based Systems, vol. 138. no. 148, p. 25 (1996)
19. Noyn, F., Mc Guinness, D.L.: Ontology development 101: a guide to creating your first ontology (2001). http://protege.stanford.edu/publications
20. Uschold, M., Gruninger, M.: Ontologies: principles, methods and applications. Knowl. Eng. Rev. **11**(2), 14–17 (1996)

21. Pinto, H.S., Martins, J.P.: Ontologies: how can they be built? Knowl. Inf. Syst. **6**, 441–464 (2004)
22. Fernández-López, M., Gómez-Pérez, A., Juristo, N.: Methontology: from ontological art towards ontological engineering, no. 5, pp. 33–40 (1997)
23. The KACTUS booklet version 1. 0. esprit project 814. (2015). http://www.swi.psy.uva.nl/prj ects/NewKACTUS/Reports.html
24. Chen, P., Lu, Y., Zheng, V.W., et al.: KnowEdu: a system to construct knowledge graph for education. IEEE Access **6**, 31553–31563 (2018)
25. Youn, J., Naravane, T., Tagkopoulos, I.: Using word embeddings to learn a better food ontology. Front. Artif. Intell. **3**, 1–8 (2020)
26. Ferré, A., Deléger, L., Bossy, R., Claire, N.: C-norm: a neural approach to few-shot entity normalization. BMC Bioinform. **21**, 579 (2020)

Federated Learning-Based Driving Strategies Optimization for Intelligent Connected Vehicles

Wentao Wu[✉][iD] and Fang Fu[iD]

School of Physics and Electronic Engineering, Shanxi University,
Taiyuan, Shanxi, China
sxtwuw@163.com, fufang0621@sxu.edu.cn

Abstract. Thanks to smart manufacturing and artificial intelligence technologies, intelligent connected vehicles (ICVs) is emerged as a main transportation means. However, due to the limitations of finiteness and privacy of driving data, ICVs may not be able to share their data with other vehicles, which limits the development of ICVs. To overcome aforementioned challenges, we propose a federated learning-based driving strategy optimization scheme for ICVs. Conditional imitation learning is employed to obtain a single-vehicle intelligent driving strategy. To improve the driving ability while ensuring data privacy, federated learning is leveraged to aggregate driving policies of different ICVs. Finally, the experimental results based on the Carla platform show that the single-vehicle intelligent driving strategy achieves a high level of accuracy, and the federated learning vehicle model achieves a significant 15% increase in the success rate of turning tasks and a 21% increase in the success rate of going straight, which verifies the effectiveness of the method in this paper.

Keywords: Intelligent connected vehicles · Conditional imitation learning · Federated learning · Carla

1 Introduction

In recent years, Artificial Intelligence [1], self-driving technology development is in full swing, which provides great convenience for people's lives. However, the field of autonomous driving is currently facing two major challenges: safe driving and limited computing power. In complex and changing road conditions, Tesla and Google self-driving vehicles have been involved in serious traffic accidents [2]. At the same time, vehicles with insufficient computing power cannot accurately and quickly assess the current traffic conditions, and then there is a high probability of making wrong decisions, even leading to traffic accidents [3,4].

The emergence of imitation learning provides an opportunity to solve the problem of safe driving in single vehicle intelligence [5]. Imitation learning is characterized by fast learning speed and decision making by imitating expert

C. Yu et al. (Eds.): GPC 2022, LNCS 13744, pp. 67–80, 2023.
https://doi.org/10.1007/978-3-031-26118-3_5

strategies, for which researchers have conducted intensive research. For example, reference [6] designed an end-to-end imitation learning system that maps from sensor data to controller commands, which in turn rationally adjusts the vehicle speed as well as the driving trajectory. Reference [7] proposed an imitation learning safety framework that uses data set aggregation to reduce the amount of computation and enables the controller to quickly output commands, which improves the response speed of the vehicle. However, imitation learning tends to focus only on whether the vehicle can pass the expert demonstration and arrive at a driving strategy, which can ensure safety but can be problematic in scenarios where there are multiple driving options at intersections. Conditional imitation learning can solve this problem by adding advanced conditional commands to imitation learning, e.g., go straight at the next intersection, turn left at the next intersection. Reference [8] applied conditional imitation learning to single vehicle intelligence and trained the vehicle model by adding advanced instructions, and the tested vehicle can make accurate decisions quickly at intersections, solving the problem of confusing vehicle driving strategies at intersections. Reference [9] collected a large number of drivers' operations in the face of obstacles as advanced instructions and uses image enhancement in training the model, and the results show that the vehicle improves the accuracy of obstacle avoidance and also avoids sudden stopping of the vehicle in the process of travel. Reference [10] proposed a multitask conditional imitation learning framework to adapt to lateral longitudinal control tasks, allowing the vehicle to make safe and effective decisions at crowded intersections. Advanced conditional commands are particularly important when facing complex road conditions, and accurate and concise commands can make conditional imitation learning achieve twice the result with half the effort.

With the rapid development of communication technologies such as 5G [11], more intelligent and advanced smart connected vehicles have emerged based on the existing self-driving vehicles. Smart connected vehicles are based on sensors, controllers, and other devices with functions such as environmental awareness and cooperative control, which can achieve safe and energy-efficient smart driving [12]. Reference [13] described the current key technologies for smart connected vehicles: vehicle-road collaboration, autonomous vehicle decision making, external environment sensing, and control execution. Smart. Internet-connected vehicles are more advanced in terms of on-board devices, having one or more of visual perception, laser perception, and microwave perception sensors to provide a more comprehensive perception of the surrounding environment [14], and have a larger amount of data compared to ordinary single vehicles [15]. However, while intelligent networked vehicles bring convenience to people's transportation, the complexity of the system and the increase of external communication interfaces can make the vehicles more vulnerable to cyber attacks, and information security becomes an important aspect [16]. The emergence of federated learning provides an effective way to ensure information security, as well as the privacy of users. Federated learning has the characteristics of user privacy protection, adapting to large-scale data model training [17,18], ensuring that the data is not shared among the participants and that large-scale data is trained on the model to improve the training quality. Adding federated learning to the smart networked vehicle not only solves the lack of data volume of a single vehicle, but also solves the key problem of limited computing power of a single vehicle. To this end, this paper proposes a scheme for optimizing the driving performance of smart connected vehicles based on conditional imitation learning and federated learning.

2 Scene and Model

2.1 System Scene

We design a system scenario as shown in Fig. 1, where vehicles use local data generated while driving to train a model and upload the model parameters to a Road Side Unit (RSU), which aggregates the model parameters of multiple vehicles and distributes the aggregated results to each vehicle in the coverage area, continuously iterating to improve the overall vehicle driving performance.

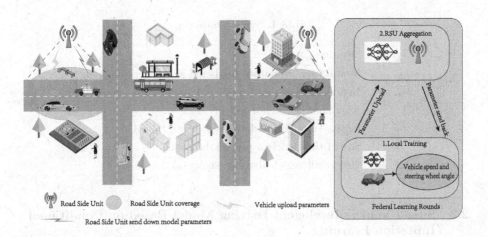

Fig. 1. System scene

With N cars, each car is indexed by n. Each car generates data in real time while driving: the photos x_n taken by the camera, the speed, throttle and brake of the vehicle at the moment corresponding to the photos together constitute the tags y_n, which in turn form a local private data set $D_n = \{x_n, y_n\}$.

The vehicle local training uses conditional imitation learning, and the resulting dataset is used for training the neural network part of conditional imitation learning, such as the neural network in Fig. 2, to continuously optimize the vehicle's driving strategy through training.

Each training sample k in the data set consists of 200 (x_n, y_n), the input of the neural network is $x_k \in [x_n]$, the output of the neural network is $y_k \in [y_n]$, the vehicle model receives the data input and gets the predicted output, the gap between the predicted output and the actual output is represented by the loss function $\ell(w)$:

$$\ell(w) = \ell(a_p, a_{real}) = \ell\left\langle (s_p, a_p), (s_{real}, a_{real}) \right\rangle = \|s_p - s_{real}\|^2 - \lambda \|a_p - a_{real}\|^2 \tag{1}$$

The loss function $L(w_n)$ after the vehicle is trained through the full private dataset is defined as:

$$L(w_n) = \frac{1}{|D_z|} \sum_{z=1}^{|D_z|} \ell(w) \tag{2}$$

After continuous iterations the model parameters are aggregated and sent down the federal learning loss function $L(w_G)$ as:

$$L(w_G) = \sum_{i=1}^{N} \frac{|D_z|}{|D|} L(w_n) \tag{3}$$

The meaning of the symbols used in the formula is shown in the Table 1.

Table 1. Symbol meaning

Symbol	Meaning
$a_p,\ a_r$	Neural network output predicted action actual action
$s_p,\ s_r$	Neural network input predicted state, actual state
λ	A number between 0 and 1
z	Iterate through the data set starting from 1
$|D_z|$	Size of the vehicle n private data set
$|D|$	Sum of all vehicle data set sizes

2.2 Single Vehicle Intelligent Driving Model Based on Conditional Imitation Learning

The principle of conditional imitation learning is shown in Fig. 3, where an additional command $c = c(n)$, command provided by the expert, represents the information that is helpful for the expert to provide the command, and can express the expert's intention, destination information, etc. For example, the vehicle expresses the intention to change lanes by turn signals, and when passing through an intersection, the destination information is used to choose to go straight or turn to reach the destination more quickly, and these turn signals and target information can be used as commands in the formula.

The training dataset becomes $D = \{\langle s_t, c_t, a_t \rangle\}_{t=1}^{T}$, and the objective of conditional imitation learning is shown in Eq. (4).

$$\underset{\theta}{\mathrm{min}}\, imize \sum_{t} \ell\left(F(s_t; c_t; \theta), a_t\right) \tag{4}$$

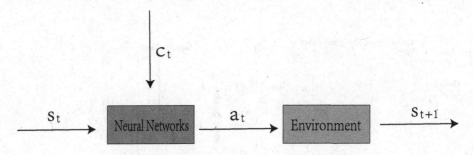

Fig. 2. Principle of conditional imitation learning

As shown in Fig. 3, the conditional imitation learning network structure is divided into image and measurement modules, where the photo module is a convolutional neural network and the measurement module is a fully connected network, using image i and a set of vectors m (corresponding to the parameters among the images) as inputs, respectively.

$$j = J(i, m) = \langle I(i), M(m) \rangle \tag{5}$$

For the command c, $C = (c(0), c(1), ..., c(k))$ each command $c(i)$ corresponds to a branch of the A^i network, each network branch A^i can only output a specific action, the command c plays a role of selecting a specific network branch, so the output of the network is:

$$F(i, m, c(i)) = A^i(J(i, m)) \tag{6}$$

Different A^i branches correspond to different sub-strategies, and each branch is equivalent to a module in a driving scenario, e.g., a module for going straight, a module for turning left, and many other scenarios. In the normal driving of the car, the picture with $200 \times 88 \times 3$ pixels observed by the camera at the current moment and the speed of the car together constitute the state s, and the action a is acceleration and steering wheel steering angle. The car model is trained several times to go straight, turn, and avoid vehicles, thus achieving the effect of single vehicle unmanned driving.

2.3   Federated Learning-Based Model Aggregation

Each participant n trains the local model with its dataset $D_{n \in N}$ and sends the local model parameters to the central node to generate the global model after aggregating the local model parameters.

As shown in Fig. 4, the initial phase of federated learning initializes the vehicle model parameters w_G^0, followed by updating the local model parameters w_n^t using local data training for each vehicle based on the global model w_G^t, t denotes the current iteration index, the vehicle n goal is to minimize the loss function $L(w_n^t)$, and finally the updated local model is uploaded to the server.

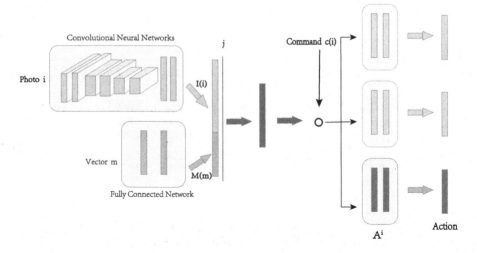

Fig. 3. Network structure of conditional imitation learning

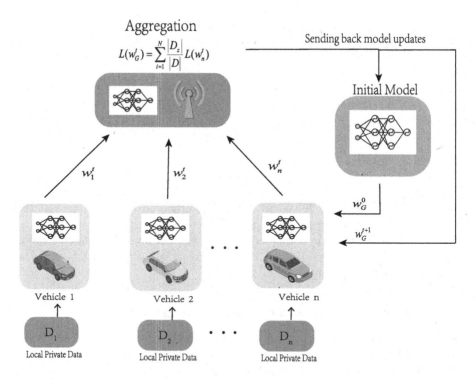

Fig. 4. Federated learning schematic

The roadside unit aggregates the participants' local models, distributes the updated global model parameters w_G^{t+1} to the participants, and the server minimizes the global loss function $L(w_G^t)$:

$$L(w_G^t) = \frac{1}{N} \sum_{n=1}^{N} L(w_n^t) \tag{7}$$

The above process is continuously iterated to minimize the global loss function $L(w_G^t)$.

In the whole federated learning process, because each vehicle's database is different, resulting in some vehicles driving better in straight scenarios, while some vehicles driving better in turning scenarios, each vehicle local model parameters have differences, and different model parameters in the federated learning aggregation, the weight of the size of the aggregated model will directly affect the effect of the aggregated model.

For this reason, we use the federated averaging algorithm to solve this problem. Each vehicle model parameter has the same weight in the aggregation process, so that different vehicles can jointly improve the overall networked vehicle driving performance, using the Federated Averaging Algorithm [18].

Algorithm 1: Federated Averaging Algorithm

Require: Local minibatch size B, Local minibatch size B, number of participants m per iteration, number of local epochs E, and learning rate η
Ensure: Global model W_G
[Participant i](Local Vehicles)
LocalTraining (i,w)
Split local dataset D_i to minibatches of size B which are included into the set B_i
for *each local epoch $j = 1 to j = E$* **do**
  for *each $b \in B_i$* **do**
    | $w \leftarrow w - \eta \Delta L(w; b)$ (ΔL is the gradient of L on b)
  end
end
[Server]
Initialize w_G^0
for *each iteration $t = 1$ to $t = T$* **do**
  Randomly choose a subset S_t of m participants from N
  for *each partipant $i \in S_t$* **do**
    | $w_i^t + 1 \leftarrow$ LocalTraining (i,w)
  end

  $w_G^t = 1/\sum_{i \in N} D_i \sum_{i=1}^{N} D_i w_i^t$ (Average aggregation)
end

---

3   Simulation Verification

3.1   Carla Simulation Platform

We use the Carla simulation platform for dynamic validation and testing of the model, Carla [19] is an open source simulator for urban driving that supports training, prototyping and validation of autonomous driving models, including perception and control. The Carla platform used, version 0.8.3, provides two professionally designed towns as shown in Fig. 5.

(a) Town 1 (training)

(b) Town 2 (test)

Fig. 5. Town map provided by Carla platform

3.2   Database Preprocessing

The used public dataset [8] includes training set and test set, which consists of 657800 photos and their corresponding labels, and contains four scenarios along the road straight ahead, left turn, intersection straight ahead, and right turn, which account for 34%, 23%, 20%, and 23% in the dataset, based on the large dataset can ignore the influence of different percentages of the four scenarios on the model training.

The full training set was used for training the bicycle intelligence, and at the end of the training, the model was tested on the test set, which led to the loss function.

In federated learning, the training set is divided into 80% of the data with uniform scene share for model pre-training, and the remaining 20% of the data are equally divided for further training of the vehicle.

3.3   Simulation Parameter Setting

The training of the single vehicle intelligence model and the federated learning of all participating vehicle models are trained using neural networks, and the neural network settings are shown in Table 2.

The initial learning rate is set to $a = 0.0001$, and the batch is set to 200 for training single vehicle intelligence. For the federated learning training, the batch is set to 120, and the instructions contain codes for four scenarios: drive along the road (0), go straight at the intersection (1), turn left (2), and turn right (3), where 0 is the default instruction. In the federated learning process, the number of local vehicles is set to 3, and each vehicle is trained using 1/3 of the remaining data set excluding the pre-trained model.

Table 2. Neural network settings [19]

Module	Input dimension	Channels	Kernel	Stride
Perception	$200 \times 88 \times 3$	32	5×5	2
	$200 \times 88 \times 3$	32	5×5	2
	$98 \times 48 \times 32$	32	3×3	1
	$96 \times 46 \times 32$	64	3×3	2
	$47 \times 22 \times 64$	64	3×3	1
	$45 \times 20 \times 64$	128	3×3	2
	$22 \times 9 \times 128$	128	3×3	1
	$20 \times 7 \times 128$	256	3×3	2
	$9 \times 3 \times 256$	256	3×3	1
	$7 \times 1 \times 256$	512	–	–
	512	512	–	–
Measurement	1	128	–	–
	128	128	–	–
	128	128	–	–
Joint input	512+128	512	–	–
Control	512	256	–	–
	256	256	–	–

3.4   Analysis of Simulation Results

The loss function is shown in Fig. 6, which shows the loss function in the training process and the loss function in the test set. The shaded part of the figure shows the maximum value as well as the minimum value of both in the iterative process. After training up to 450000 times, the decreasing trend of training loss gradually slows down and the testing loss tends to be constant. To prevent model overfitting, the model trained up to 490000 times was selected for dynamic validation after several tests.

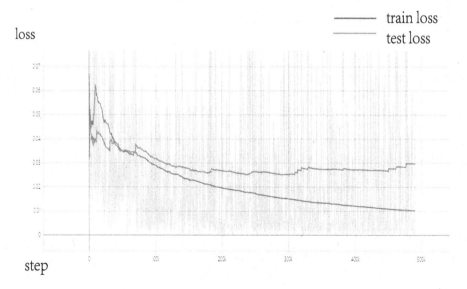

Fig. 6. Single vehicle intelligence loss function

After training the model, we simulated the model dynamically on the Carla platform, and the experimental results statistics are shown in Fig. 7(a).

The model uses town 1 for training and town 2 for testing, which leads to a small difference in the success rate of the two tasks. Both perform well in the straight ahead scenario, with a task success rate of 89%, and similarly both perform similarly in the turning scenario, which shows the reasonableness and effectiveness of conditional imitation learning applied to the intelligent driving of vehicles.

Federated learning model pre-training used 80% of the training set, pre-training model trained 94000 steps, compared with the bicycle intelligence, the training set use becomes less, the number of training reduced, making the federated learning pre-training model compared with the bicycle intelligence in the dynamic verification effect, using the federated learning subsequent training of 56000 steps, a total of 1500000 steps, under the test set experimental results are shown in Fig. 7(b).

According to the Fig. 7(b), we can see that the model effect is improved very obviously after federated learning, the success rate of intersection turning is improved from 15% to 30%, and the success rate of straight driving is improved from 67% to 88%. This shows that by federated learning, the vehicle driving strategy can be optimized and the safety is improved at the same time.

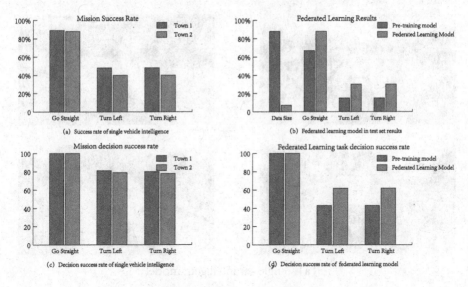

Fig. 7. Experimental result

At the same time the success rate of correct vehicle decision is equally impor-tant, as shown in Fig. 7(c), the single vehicle intelligent model in the straight scenario are 100% complete correct decision, in the turn reached about 80% success rate, the training and testing gap is very small, thus verifying that the single vehicle intelligent model effect is better and universal. The reasons for the difference between the vehicle decision success rate and the mission success rate are: the vehicle speed is not enough, and the correct decision is made in the turn but the turn is not successful, which are all within the reasonable error range.

As shown in Fig. 7(d), there is also a significant improvement in the success rate of decision making for vehicles after federated learning, with the success rate of turning decision increasing from 43% to 62%, thus highlighting the effec-tiveness and rationality of federated learning.

The single vehicle intelligence model and the federated learning model are demonstrated on the Carla platform as shown in Fig. 8.

(a) Single car intelligent model

(b) Federated Learning Model

Fig. 8. Dynamic demonstration of trained vehicle intelligence model and federated learning model in Carla simulation platform

4   Summary and Outlook

In this paper, we use conditional imitation learning to implement single-vehicle intelligence driving and dynamically validate it on the Carla platform, obtaining more satisfactory results. Based on this, federated learning is used to extend single-vehicle intelligence to smart connected vehicles. The dynamic validation results show that federated learning significantly optimizes the driving strategies of vehicles participating in federated aggregation.

With the rapid development of the times, the speed of hardware and algorithm updates and iterations is also accelerating, through hardware to accelerate the speed of network computing, algorithms to strengthen the network performance, there is hope to achieve full autonomous driving as early as possible. At the same time, on the basis of achieving single-vehicle intelligence, expanding it to multi-vehicle intelligence leaves a large gap, and it is still a huge challenge to aggregate different vehicle model parameters and control multi-vehicle driving strategies.

Acknowledgements. This work was supported in part by the National Natural Science Foundation of China under Grant 62261019 and in part by the Fundamental Research Program of Shanxi Province under Grants 202103021224024 and 202103021223021.

References

1. Teoh, E.R., Kidd, D.G.: Rage against the machine? Google's self-driving cars versus human drivers. J. Saf. Res. **63**, 57–60 (2017)
2. Zhang, Z., Yu, F.R, Fu, F., et al.: Joint offloading and resource allocation in mobile edge computing systems: an actor-critic approach. In: 18th IEEE Global Communications Conference, Abu Dhabi, UAE (2018)
3. Gambi, A., Huynh, T., Fraser, G.: Generating effective test cases for self-driving cars from police reports. In: Proceedings of the 2019 27th ACM Joint Meeting on European Software Engineering Conference and Symposium on the Foundations of Software Engineering, pp. 257–267. FSE, Tallinn (2019)
4. Du, J., Cheng, W., Lu, G., et al.: Resource pricing and allocation in MEC enabled blockchain systems: an A3C deep reinforcement learning approach. IEEE Trans. Netw. Sci. Eng. **9**, 33–44 (2022)
5. Le Mero, L., Yi, D., Dianati, M., et al.: A survey on imitation learning techniques for end-to-end autonomous vehicles. IEEE Trans. Intell. Transp. Syst. **4**(1), 1–20 (2022)
6. Pan, Y., Cheng, C.A., Saigol, K., et al.: Agile autonomous driving using end-to-end deep imitation learning. arXiv preprint arXiv:1709.07174 (2017)
7. Sun, L., Peng, C., Zhan, W., et al.: A fast integrated planning and control framework for autonomous driving via imitation learning. In: Dynamic Systems and Control Conference, pp. V003T37A012 (2018)
8. Codevilla, F., Müller, M., López, A., et al.: End-to-end driving via conditional imitation learning. In: 2018 IEEE International Conference on Robotics and Automation, pp. 4693–4700. IEEE, Brisbane (2018)
9. Zhang, E., Zhou, H., Ding, Y., et al. : Learning how to avoiding obstacles for end-to-end driving with conditional imitation learning. In: Proceedings of the 2019 2nd International Conference on Signal Processing and Machine Learning, pp. 108–113. ACM, USA (2019)
10. Zhu, Z., Zhao, H.: Multi-task conditional imitation learning for autonomous navigation at crowded intersections. arXiv preprint arXiv:2202.10124 (2022)

11. Zhang, Z., Wang, R., Yu, F.R., et al.: QoS aware transcoding for live streaming in edge-clouds aided HetNets: an enhanced actor-critic learning approach. IEEE Trans. Veh. Technol. **68**(11), 11295–11308 (2019)

12. Kuang, X., Zhao, F., Hao, H., et al.: Assessing the socioeconomic impacts of intelligent connected vehicles in China: a cost-benefit analysis. Sustainability **11**(12), 3273 (2019)

13. Deren, L.I., Yong, H., Mi, W., et al.: What can surveying and remote sensing do for intelligent driving? Acta Geodaetica Cartograph. Sinica **50**(11), 1421 (2021)

14. Yang, D.G., et al.: Intelligent and connected vehicles: current status and future perspectives. Sci. China Technol. Sci. **61**(10), 1446–1471 (2018). https://doi.org/10.1007/s11431-017-9338-1

15. Fu, F., Kang, Y., Zhang, Z.: Soft actor-critic DRL for live transcoding and streaming in vehicular fog computing-enabled IoV. IEEE Internet Things J. **8**(3), 1308–1321 (2020)

16. Jadhav, S., Kshirsagar, D.: A survey on security in automotive networks. In: International Conference on Computing Communication Control and Automation, pp. 1324–1330. IEEE, Shanghai (2018)

17. Zhang, Z., Zhang, Q., Miao, J., et al.: Energy-efficient secure video streaming in UAV-enabled wireless networks: a safe-DQN approach. IEEE Trans. Green Commun. Netw. **5**(4), 1982–1995 (2021)

18. Zhang, C., Xie, Y., Bai, H., et al.: A survey on federated learning. Knowl.-Based Syst. **216**, 106775 (2021)

19. Dosovitskiy, A., Ros, G., Codevilla, F., et al.: CARLA: an open urban driving simulator. In: Conference on Robot Learning, pp. 1–16. PMLR, New York (2017)

Traffic Sign Image Segmentation Algorithm Based on Improved Spatio-Temporal Map Convolution

Qianying Zou[1] , Lin Xiao[2], Guang Xu[2], Xiaofang Wang[1(✉)] , and Nan Mu[3]

[1] Geely University of China, Chengdu 641423, Sichuan, China
939549393@qq.com
[2] Chengdu College of University of Electronic Science and Technology of China,
Chengdu 611731, Sichuan, China
[3] Sichuan Normal University, Chengdu 610066, Sichuan, China

Abstract. A traffic sign image segmentation algorithm based on improved spatio-temporal graph convolution is proposed by fusing octave convolution and spatio-temporal component graph convolution network for the problem of road traffic sign recognition in a complex environment. The algorithm uses octave convolution to reduce the computational effort to improve the recognition speed and uses a spatio-temporal graph convolution network to recognize traffic signs more accurately. First, the acquired images are processed by data image enhancement; then, the RGB image saliency detection module based on octave convolution is used; then, the spatio-temporal map convolution network module is improved by using the SETR algorithm to train a lightweight and high-precision spatio-temporal map convolution network model; finally, the image details are optimized by using the octave convolution residual module and eventually used for road sign recognition. The experimental results show that the algorithm can effectively improve the segmentation accuracy rate, which is 16.5%, 10.1%, 6.1%, and 5.1%, respectively, compared with other algorithms; in terms of recognition speed, the single image processing time of different data sets is better than other algorithms; its recognition effect is also better than other algorithms in terrible weather conditions, such as intense light, fog, heavy rain, night and snow conditions, especially in intense light, heavy rain and low contrast weather conditions. In the ablation comparison experiments, its algorithm improves 12.5%, 7.3%, and 8.6% in segmentation accuracy compared with other module combination algorithms in the same data set case.

Keywords: Intelligent traffic · Image segmentation · Spatio-temporal map convolution · Traffic signs · Octave convolution · Complex environment

1 Introduction

Traffic sign recognition is an essential part of the unmanned system for real-time road navigation. Its recognition accuracy and speed will directly affect

C. Yu et al. (Eds.): GPC 2022, LNCS 13744, pp. 81–98, 2023.
https://doi.org/10.1007/978-3-031-26118-3_6

the safety of the unmanned system [1]. However, there is still room for further improvement of traffic sign recognition in complex environments, such as the correct recognition rate in terrible weather and the recognition speed during vehicle driving [2].

Deng Xiangyu [3] et al. proposed a shape recognition algorithm for traffic sign classification combined with BP neural network. The algorithm uses color information to segment the traffic sign area. Still, the algorithm has a specific false recognition rate for circles and square octagons and is vulnerable to bad weather and road congestion, resulting in a low recognition rate. Xu Jingcheng [4] et al. proposed an improved traffic sign recognition method based on the Alex model by introducing a batch normalization method and adding a global average pooling layer to reduce the depth of the network. Still, the algorithm needs to be improved to achieve recognition processing of multiple complex traffic signs in realistic traffic environments. For complex environments, He Ruibo [5] et al. proposed an improved deep learning algorithm for road traffic sign recognition, combining SENet model and ResNet model, extracting the advantages of each, and using a smaller number of network layers to achieve a higher recognition level, while the algorithm has more manual intervention and more neural network parameters adjustment and input. Dewi [6] et al. used the SPP concept to improve Yolo V3, Resnet 50, Densenet, and Tiny Yolo V3 backbone networks for constructing traffic sign feature extraction, but the algorithm has high computational complexity and cumbersome implementation process. Cao [7] et al. proposed an intelligent vehicle traffic sign detection and recognition algorithm with an improved LeNet-5 convolutional neural network model to solve the problem that traditional traffic sign recognition is easily affected by environmental factors, however, the traffic sign recognition method based on deep learning is computationally intensive and has poor real-time performance. Yazdan et al [8]. proposed a shape classification algorithm based on an SVM classifier to improve the segmentation accuracy and filter the wrong pixels in the classification result by symbolic geometry. However, it is time-consuming and cannot meet the real-time requirements of vehicles in the driving process. Di Lan et al. [9] et al. proposed a road traffic sign recognition algorithm based on a possibility clustering algorithm and convolutional neural network, mainly to solve the high time consumption caused by noise and complex backgrounds in images for picture recognition. Still, the algorithm has certain errors in traffic sign recognition. [12] Jie Wei [10] et al. proposed a real-time traffic sign recognition method based on multi-feature fusion, which mainly addresses the impact of poor real-time performance due to the difference in sample categories in the recognition process. Mannan [11] et al. proposed a completely data-driven segmentation technique to solve the problem of complete separation of the corresponding pixels of traffic signs from the background objects. Still, this method comes at the cost of increasing the computational cost. Handoko [12] implements traffic sign color and shape segmentation based on a reduced algorithm operation cost, but there is a specific error in terrible weather. James [13] et al. proposed a capsule-based neural network to replace the commonly used CNN and RNN. However, only in the Indian traffic dataset, the assessment accuracy has been improved to a certain extent, and the strong locality does not have universal applicability.

Based on the problems above, the study proposes a traffic sign image segmentation algorithm based on improved spatio-temporal map convolution. Firstly, the algorithm performs data image enhancement on the acquired images and then inputs the RGB image saliency detection module that uses octave convolution instead of normal convolution to improve the computing efficiency to form a preliminary feature map. Then, the initial feature map is stitched with the output feature map after spatio-temporal map convolution and fed into the SETR (Segmentation Transformer) algorithm for feature matching to improve the correct segmentation rate. Finally, the feature maps are fed into the Octave Convolution Residual module for edge optimization to reduce the computational effort to enhance the recognition speed.

2 Related Job

2.1 Pre-processing

Traffic sign image segmentation is often complex due to lousy weather, illumination and other aspects. The study uses the IPT (Image Processing Transformer) model to pre-process the images and accomplish image enhancement such as super-resolution and denoising to reduce the influence of environmental factors on traffic sign recognition in complex scenes [14].

IPT is a pre-training model for end-to-end image processing consisting of multiple head and tail structures processing different tasks and a single shared body, with a framework of multiple head structures, encoders, decoders and multiple tail structures.

During the IPT pre-processing, the multi-head structure processes the image into feature maps, as shown in Eqs. 1 and 2.

$$f_H = H^i(x) \tag{1}$$

$$f_H \in R^{C \times H \times W} \tag{2}$$

where H^i $(i = \{1, \ldots, N_t\})$ denotes the ith task header and Nt denotes the number of tasks, i.e., the size of the input dataset. The multi-head structure generates a feature map $f_H \in R^{C \times H \times W}$ with C channels and the same width and height, and C is usually 64. The feature map is cut and stretched by the feature map operation, and the feature map is cut into N pieces according to the size P × P. Each feature piece is spread into a dimensional $P^2 \times C$ vector to obtain the cut part, as shown in Eq. 3.

$$f_{p_i} \in R^{P^2 \times C}, i = \{1, \ldots, N\} \tag{3}$$

where $f_{(p_i)}$ is the feature vector that has been cut and levelled. It is fed into a Transformer for processing, as shown in Eq. 4.

$$f_{D_i} \in R^{P^2 \times C} \qquad\qquad (4)$$

where f_{D_i} is the same dimensional output feature obtained after Transformer processing. Finally, the features f_{D_i} are fed into the multi-tailed structure for dimensional transformation and decoded into the target image.

2.2 Octave Convolution

Traffic sign images in complex scenes can be decomposed into low-frequency and high-frequency signals. The high-frequency signal represents the rich details of the image with drastic changes and large grayscale differences between adjacent areas. The low-frequency signal represents the gently changing edge structure with a slowly changing grayscale.

In traffic sign segmentation, to improve the segmentation accuracy and reduce the computational effort, the study uses octave convolution instead of a full convolution module and a common convolution module in the residual module in the spatio-temporal component map convolution algorithm. Octave convolution can effectively improve the image segmentation accuracy, solve the problem of spatial redundancy in the convolutional computation process, and realize the lightweight network architecture design [15].

Classify and convolve the feature maps, as shown in formulas 5 and 6.

$$Y^{H \to H} = f_1 (X_h) \qquad\qquad (5)$$

$$Y^{L \to L} = f_2 (X_l) \qquad\qquad (6)$$

where X_h denotes the high-frequency component, X_l denotes the low-frequency component, and $f(\bullet)$ denotes the convolution operation. The process of low-frequency component to high-frequency output is first to perform convolution and upsampling on the low-frequency component X_l The process of low-frequency component to high-frequency output is first to convolve and upsample the low-frequency component, and restore its resolution to the same as the high-frequency component, as shown in Eq. 7. Similarly, the process from the high-frequency component to the low-frequency output is to downsample and convolve the X_h down-sampling operation and convolution, as shown in Eq. 8.

$$Y^{L \to H} = upsample (f_3 (X_l)) \qquad\qquad (7)$$

$$Y^{H \to L} = f_4 (pool (X_h)) \qquad\qquad (8)$$

$pool(\bullet)$ denotes the downsampling operation, and $upsample(\bullet)$ denotes the upsampling operation. The number of input channels of the convolution layer f_3 is equal to the number of channels of the low-frequency component, and the number of output channels is equal to the number of channels of the high-frequency component. The number of input and output channels of the convolution layer f_4 is opposite to that of f_3, to keep the uniform number of channels and the superposition of features between high and low frequencies, as shown in Eqs. 9 and 10.

$$Y_l = Y^{L \to L} + Y^{H \to L} \tag{9}$$

$$Y_h = Y^{H \to H} + Y^{L \to H} \tag{10}$$

where Y^m (m stands for $H \to H$, $L \to H$, and $H \to L$, $L \to L$) is the result calculated by Eq. 5, Eq. 6, Eq. 7 and Eq. 8.

2.3 Spatio-Temporal Component Map Convolution

The application of spatio-temporal component graph convolution in traffic was first proposed for traffic flow prediction [11], a structured graph representation model based on components or nodes [16]. This convolution treats each $1 \times 1 \times C$ dense grid as an image feature for simplifying and improving the efficiency of operations.

The model first constructs a spatio-temporal component map. The feature map $\{\hat{z}_{t-K}, \ldots, \hat{z}_{t-1}\}$ in which each $1 \times 1 \times C_1$ dense grid is considered as an image feature component, where t represents the ordinal number of the input image and K represents a positive number. To represent the spatio-temporal target model, with $N = h \times w$ component nodes and $K(i.e., t - K, ..., t - 1)$ on the time sequence to construct an undirected spatio-temporal component graph, as shown in Eq. 11.

$$G_{ST} = (V, E) \tag{11}$$

where V and E are the sets of nodes and edges in the undirected graph, the set of nodes $V = \{v_{kn} \mid k = t - 1, \ldots, t - K; n = 1, \ldots, N\}$ contains all nodes in K, and $F(v_k n)$ is the feature vector. The edge set E contains two types of edges: the first type is the space edge $E_S = \{v_{ki}v_{kj} \mid 1 \leq i, j \leq N, i \neq j\}$, which represents the relationship between nodes within each image feature, and since the features in the images will have various changes with time, the study uses fully connected graphs to describe the spatial relationships; the second class is the temporal edges $E_T = \{v_{ki}v_{(k+1)j}\}$, which represents the relationship between nodes of similar feature images, connects parts or nodes with the same position in the similar image features, which can be regarded as a tracking trajectory of a specific part transformed over time.

Based on the above basis, the study uses a graph convolutional network to process the relationship between its nodes. First, the two-layer graph convolutional network is used to output the matrix $\{\hat{z}_{t-K}, \ldots, \hat{z}_{t-1}\}$, (where $\hat{Z}_{t-1} \in R^{h \times w \times C_2}$), an undirected spatial component graph is reconstructed using maximum pooling aggregated spatio-temporal component features G_S. Then, the spatio-temporal part graph features G_ST and spatial part features G_S are aligned and stitched together as a whole, and the global convolution module is used to match the features of these two parts. In this module, all the convolutional layers produce 256 channels of the feature map, and the output features are denoted as Z.

The output features of the spatio-temporal component feature model and the spatial component feature model have different characteristics. An attention mechanism assigns different weights to all features, i.e., feature channel selection. The feature Z is transformed by nonlinear transformation as $\hat{Z}$, as shown in Eq. 12 and Eq. 13.

$$\hat{Z} = Z \otimes W \tag{12}$$

$$W = \varphi\left(\theta_2 \psi\left(\theta_1 f_{GAP}(Z)\right)\right) \tag{13}$$

where $\otimes$, θ, ψ, f_{GAP} denote the channel-by-channel multiplication, the sigmoid activation function, the ReLU activation function and the global average pooling, respectively. θ_1 and θ_2 are the convolution layer weights.

Finally, the features are $\hat{Z}$ are input to the decoding module and connected with the input image features Z in the coding model to obtain the output target image.

3   Improved Algorithm

3.1   Framework of Traffic Sign Recognition Algorithm Based on Improved Spatio-Temporal Convolution

Since the RGB saliency detection module cannot capture image information quickly and accurately in extreme cases, it can significantly improve the efficiency of saliency detection operations after introducing octave convolution for optimization. The traditional spatio-temporal component map convolution module has high recognition efficiency only for small datasets, and the recognition rate is poor in the case of larger datasets, so the algorithm utilizes the SETR model to improve the recognition rate. Finally, the conventional ordinary convolution is replaced by octave convolution in the improved residual module to enhance the edge optimization of traffic sign images. The algorithm improves the segmentation accuracy and recognition speed of traffic signs in complex scenes from four aspects, as shown in Fig. 1.

As shown in Fig. 1, the core flow of the algorithm is as follows:

1. Pre-processing of traffic sign image datasets for image enhancement in complex scenes.
2. Performing RGB saliency detection on pre-processed images and replacing the normal convolution layer with octave convolution to enhance detection efficiency to form a preliminary feature map.
3. The initial features are fed into the spatio-temporal graph convolutional network. The acquired output features are spliced with the initial features to achieve feature matching by using the SETR algorithm to improve the correct segmentation rate.

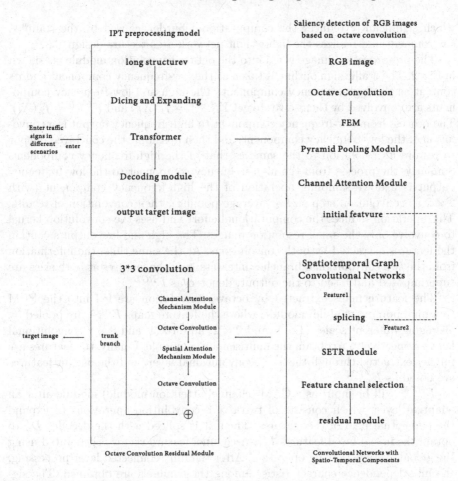

Fig. 1. Framework of traffic sign recognition algorithm with improved spatio-temporal convolution

4. The feature map is sent to the octave convolution residual module for detailing and edge optimization to reduce the computational effort to enhance the recognition speed and output the segmented image.

3.2 Octave Convolution-Based RGB Image Saliency Detection [17] Module

Environmental factors bring sizeable negative impacts on traffic sign acquisition. Still, the color and structure features of traffic signs are more prominent, so RGB images can be directly used as the study object. In extreme cases, the accurate detection of important targets will increase the computational burden and reduce the computational efficiency. Therefore, the study uses octave convolution instead of the regular convolution module in the original algorithm, which significantly reduces the computational burden caused by the complex environment and realizes the lightweight network architecture design.

The pre-processed image is fed into the octave convolution module, as shown in Fig. 2. The traffic sign outline is taken as the low-frequency component and its content as the high-frequency component. The high and low-frequency components are convolved by themselves to get $Y^{H \rightarrow H} = f_1(X_h)$ and $Y^{L \rightarrow L} = f_2(X_l)$. The process from low-frequency component to high-frequency output is a convolution of the low-frequency component, and then upsample the convolved output to restore its resolution to the same as that of the high-frequency component. Similarly, the process from the high-frequency component to the low-frequency output is a down-sampling operation of the high-frequency component (with 2×2 as convolution step size or average pooling), then convolution of results. This operation reduces the computational effort and uses the convolution kernel to convolve only the small resolution images. The high and low outputs contain the features extracted by both convolutions. At the same time, the information from the two convolutional branches interacts, and the features at both sizes are superimposed and fused to the output, denoted as F^{RGB}.

The feature maps extracted by octave convolution are fed into the PPM (pyramid pooling module) module, where the feature maps F^{RGB} are pooled by averaging different scales (1×1, and 2×2, and 4×4, and 8×8 convolutional layers respectively) to obtain the multisemantic module. Finally, the features are further extracted through the 3×3 convolutional layer, and the output features are noted as F^{PPM}_{out}.

F^{PPM}_{out} will be input as a CAM (channel attention module) module after an adaptive layer, which consists of two 3×3 convolution operations to expand the perceptual field of the features. Then, it is spliced with the decoder D_n to obtain the initial fused features F^C_n. A channel feature vector is generated using the global average pooling operation. After the fully connected layer processing, the interdependence characteristics among the channels are obtained. The sigmoid function is used to weigh the importance of each channel W, so that it is multiplied with F^C_n to get the weighted feature $\hat{F}^C_n$. Finally, the w1 $\times$ 1 convolution layer is used to reduce the $\hat{F}^C_n$ The number of feature channels is reduced to the input feature size as the output of the CAM module, which is denoted as F^{CAM}_{out}.

3.3 Spatio-Temporal Component Map Convolutional Network

The spatio-temporal component graph convolution includes a spatial graph convolution and a temporal graph convolution, where the spatial graph convolution is the core.

The initial features output by the RGB image saliency detection module are firstly obtained by the spatio-temporal map convolution network as feature 1, and secondly, the F^{CAM}_{out} which are not processed by the spatio-temporal map

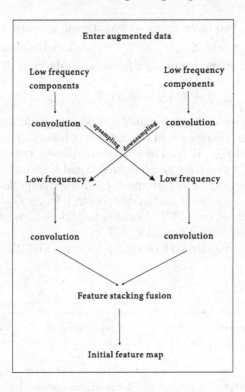

Fig. 2. Octave convolution module

convolution network as feature 2. The two are aligned in the channel and then stitched together.

The model constructs a spatio-temporal component map based on the input feature 1 $G_{ST} = (V, E)$. Next, the weights of proximity matrix A are determined based on Figure G_{ST} relations, as shown in Eq. 14 and Eq. 15.

$$\hat{A} = A + I \tag{14}$$

$$D_{\ddot{u}} = \sum_i \hat{A}_{ij} \tag{15}$$

where I is the unit matrix, $D_{\ddot{u}}$ is the sum of the weights of the adjacent matrix. The proximity matrix and the feature matrix $H^{(0)}$ are expressed as the graph convolutional network inputs, the output of the graph convolutional network is updated as $H^{(l+1)}$, as shown in Eq. 16.

$$H^{(l+1)} = \delta \left(D^{-1/2} \hat{A} D^{-1/2} H^{(l)} \Theta^{(l)} \right) \tag{16}$$

where $l = 0, 1, \ldots, l - 1$. Θ is the weight matrix of the specific layer to be trained, and δ is the nonlinear activation function ReLU.

Then, using a two-layer graph convolutional network to output matrix $\left\{\hat{Z}_{t-K}, \cdots, \hat{Z}_{t-1}\right\}$, where $\hat{Z}_{t-1} \in R^{h \times w \times C_2}$, utilizes the maximum pooling to aggregate spatio-temporal component features, as shown in Eq. 17.

$$\mathbf{z}_{ST} = MaxPooling\left(\left[\hat{\mathbf{z}}_{t-K} \cdots {}_{t-1}\right]\right) \tag{17}$$

Reconstructing an undirected spatial component graph G_S. G_S is similar to the graph G_{ST}, the difference is in the number of images. The image number of G_S is 1 and that of G_{ST} is K. The spatial component features $Z_s \in R^{h \times w \times C_2}$ are obtained using a two-layer graph convolutional network.

Next, channel alignment is performed. Combine the spatio-temporal component graph features G_{ST} with spatial component features G_S. The full convolution module is replaced by SETR algorithm to achieve high-efficiency operation and improve the recognition accuracy.

Feature 1 is feature-matched to feature 2 using the SETR algorithm, as shown in Fig. 3.

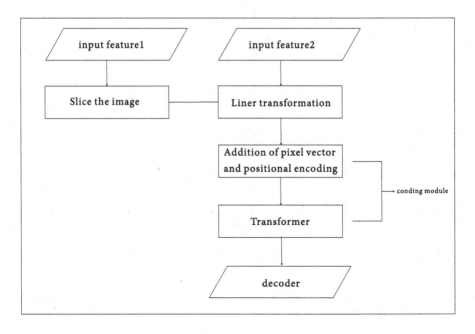

Fig. 3. SETR algorithm steps

First, convert the two-dimensional image $H \times W$ into a one-dimensional sequence. Transformer accepts the embedded $Z \in \mathbf{R}^{L \times C}$ one-dimensional sequence as input, where L is the sequence length and C is the hidden channel size; this is used to realize the input image $x \in \mathbf{R}^{H \times W \times 3}$ to Z for image serialization.

In general, the serialized images are first turned into a one-dimensional vector before image segmentation. The image is sliced into image blocks of the same size as the feature map, and all image blocks are vectorized to form a sequence as Transformer input sequence L.

Each serialized image block patch is further mapped to a potential c-dimensional embedding space using the linear mapping f (Linear Projection): $p \rightarrow e \in R^c$, where p is an image block, e is a sequence of image block compositions, e has C sequences, and e is a sequential image of a one-dimensional block embedding.

Encoding the image block patch spatial information, each position i learns a specific embedding p and adds it to e_i to form the final sequence input $E = e_1 + p_1, e_2 + p_2, \ldots, e_L+$, where L is the input sequence. Using this approach, the spatial information is preserved, as shown in Eqs. 18, 19, and 20.

$$query = Z^{l-1} W_Q \tag{18}$$

$$key = Z^{l-1} W_K \tag{19}$$

$$value = Z^{l-1} W_V \tag{20}$$

where $W_Q, W_K, W_V \in R^{C \times d}$ are the three linear projection layer learnable parameters, and d is the (query, key, value) dimensionality. The Transformer encoder comprises a multilayer self-attention mechanism (MSA) and a multilayer perceptron (MLP) block.

The self-attention mechanism (SA) is expressed formulated in Eq. 21.

$$SA\left(Z^{l-1}\right) = Z^{l-1} + softmax\left(\frac{Z^{l-1} W_Q \left(z W_K\right)^{\top}}{\sqrt{d}}\right)\left(Z^{l-1} W_V\right) \tag{21}$$

where d is usually set to C/m, $Z^{l-1} \in R^{c \times L}$.

The MSA is an extension with m independent SA operations, which are projected in series output as shown in Eq. 22.

$$MSA\left(Z^{l-1}\right) = \left[SA_1\left(Z^{l-1}\right); SA_2\left(Z^{l-1}\right); \ldots; SA_m\left(Z^{l-1}\right)\right] W_O \tag{22}$$

where $W_O \in R^{md \times C}$. The MSA output is converted by a residual MLP block and the layer output is shown in Eq. 23.

$$Z^l = MSA\left(Z^{l-1}\right) + MLP\left(MSA\left(Z^{l-1}\right)\right) \in R^{L \times C} \tag{23}$$

Finally, three decoders are designed to perform pixel-level image segmentation in the SETR algorithm to improve the segmentation accuracy.

3.4 Octave Convolution Residual Module

The study uses octave convolution to replace the full convolution in the original residual module to optimize it and merge the features in the channel attention mechanism to reduce the computational effort and improve the speed of traffic sign recognition.

The output features of the spatio-temporal component map convolutional network are denoted as $M_{i,c}(x)$, which are sent to the 3×3 convolutional layer, and its features enter the trunk branch and the CBAM (Convolutional Block Attention Mechanism) branch, respectively. The features entering the CBAM branch go through the channel attention mechanism module to construct the importance of feature channels, enhance or suppress different channels to generate weights for different tasks, identify the interdependencies between channel features and find useful features, as shown in Eq. 24.

$$M_c(F) = \sigma(MLP(AvgPool(F)) + MLP(MaxPool(F))) \tag{24}$$

where σ denotes the sigmoid activation function, MLP denotes the multilayer perceptron, and M_c represents the features entering the attention mechanism module of the channel. The features F processed by the channel attention mechanism module are multiplied with the convolved features after octave to obtain the output features, denoted as F'. F' enters the spatial attention mechanism module, which will refine the region of interest, as shown in Eq. 25.

$$M_S(F') = \sigma\left(f^{7 \times 7}\left(\begin{bmatrix} AvgPool(F') \\ MaxPool(F') \end{bmatrix}\right)\right) \tag{25}$$

where σ denotes the Sigmoid function, and $f^{7 \times 7}$ denotes the filter size of 7×7 for the convolution operation.

Multiply with the corresponding elements of F' to obtain the final output features of the CBAM branch $C_{i,c}(x)$. The trunk branch retains the original features $M_{i,c}(x)$ and adds them to the output $C_{i,c}(x)$ obtained from the CBAM branch to obtain the final output $H_{i,c}(x)$ of the residual attention module, as shown in Eq. 26.

$$H_{i+c}(x) = C_{i+c}(x) + M_{i+c}(x) \tag{26}$$

where i represents the spatial location and c represents the index of the feature channel.

4 Experiment

4.1 Experimental Environment

The datasets used in the study are from the China Traffic Sign Detection Dataset, Chinese Traffic Sign Database and Traffic Sign Dataset, as shown in Table 1.

The main experimental frequency is 2.5 GHz, NVIDIA RTX 3090 12 GB GPU, 32 GB memory server, and a pycharm software development environment.

Table 1. Traffic sign data set

Dataset	Total sample	Test sample set
China Traffic Sign Inspection Dataset (CCTSDB)	15734 sheets	600 sheets
Chinese Traffic Sign Database (CTSD)	6164 sheets	800 sheets
Traffic sign dataset (DFG)	6758 sheets	700 sheets

4.2 Experimental Evaluation Indicators

The study used both quantitative and qualitative comparisons, where the mean absolute error (MAE) [18] was chosen for the quantitative experimental evaluation metric study. The smaller its value, the better, as shown in Eq. 27.

$$\mathrm{MAE} = \frac{1}{mn} \sum_{i=1}^{m} \sum_{j=1}^{n} |\hat{y}_{ij} - y_{ij}| \tag{27}$$

The Root-Mean-Square Error (RMSE) [18] indicates the squared error expectation. The smaller the value, the smaller the error, as shown in Eq. 28.

$$RMSE = \sqrt{\frac{1}{mn} \sum_{i=1}^{m} \sum_{j=1}^{n} (y_{ij} - \hat{y}_{ij})^2} \tag{28}$$

Mean absolute percentage error (MAPE) [18], the smaller its value, the better the accuracy of the prediction model, as shown in Eq. 29.

$$MAPE = \frac{100\%}{m} \sum_{i=1}^{m} \sum_{j=1}^{n} \left| \frac{\hat{y}_{ij} - y_{ij}}{y_{ij}} \right| \tag{29}$$

where m and n represent the length and width of the image, respectively. $\hat{y}$ denotes the algorithm segmented image, and y denotes the manually segmented image.

The maximum F-measure (F-measure) [19] is a comprehensive evaluation index, and the larger its value, the more effective the experimental method, as shown in Eq. 30.

$$F_y = \frac{[(1 + y^2) * precision * recall]}{y^2 * precision * recall} * 100\% \tag{30}$$

where y2 is defined as 0.3, and the recall is the percentage of the number of relevant images detected by the algorithm and the number of all relevant images; the higher the recall, the more relevant images are segmented, as shown in Eq. 31.

$$recall = \frac{sum(S, A)}{sum(A)} \tag{31}$$

Precision [20] is the percentage of the number of segmented related images and the number of all images; the higher the precision, the more accurate the segmented related images, as shown in Eq. 32.

$$precision = \frac{sum(S, A)}{sum(S)} \tag{32}$$

where sum(S), sum(A) are the salient and the manually segmented images, respectively, and sum(S, A) is the sum of the multiplication of the values of the corresponding pixel points of both.

4.3 Comparison Experiments

Comparison with Other Algorithms in Terms of Evaluation Metrics

Table 2. Comparison of related algorithms on Evaluation Indicators

Algorithm name	MAE	RMSE	MAPE	F-measure	Precision
Improved detection recognition algorithm for LeNet-5 model	7.332%	9.663%	12.3%	86.7%	79.2%
Shape classification algorithm based on SVM classifier	7.256%	8.965%	11.6%	79.8%	85.6%
Improving AlexNet algorithm	8.064%	8.844%	10.6%	83.8%	89.6%
Improving deep learning algorithms based on SENet and ResNet models	7.503%	8.127%	10.9%	85.9%	90.6%
Algorithm of this paper	6.003%	7.806%	8.9%	93.6%	95.7%

As shown in Table 2, the experimental results are based on the CCTSDB-400 (400 images randomly selected from the "China Traffic Sign Detection Dataset", hereinafter) images. The experimental results show that the MAE, RMSE, and MAPE values of this algorithm are smaller than those of other algorithms; the F-measure index values are improved by 6.9%, 13.8%, 9.8%, and 7.7%, respectively, compared with other algorithms; compared with other algorithms, the Precision index values are improved by 16.5%, 10.1%, 6.1%, and 5.1%, respectively. Since the smaller the value of MAE, RMSE and MAPE indexes, the less error, the higher the accuracy, the larger the value of the F-measure index, the more effective the experimental method and the clearer the experimental results, and the larger the value of Precision index, the more accurate the recognition of the target of the flag image, the experimental results show that the algorithm of this paper is significantly better than other algorithms.

Comparison with Other Algorithms in Image Processing Effect

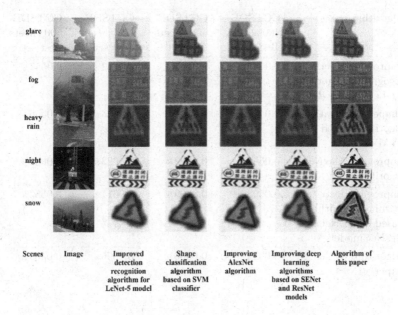

Fig. 4. Comparison of image processing effects of related algorithms

As shown in Fig. 4, qualitative comparisons with other algorithms are made, and the study selects typical cases in complex environments in the dataset for testing. The experimental results show that in the selected traffic sign images under intense light, fog, heavy rain, night and snow conditions, the proposed algorithm is significantly better than other algorithms in terms of image processing effect, especially under low contrast conditions such as heavy rain and intense light, the traffic sign images processed by the proposed algorithm are clearer.

Comparison with Other Algorithms in Image Processing Effect

As can be seen from Table 3, four different datasets, CCTSDB-400, CCTSDB-600, DFG-700 and CTSD-800, were selected to test the processing time of a single image for traffic sign recognition by different algorithms. The experiment results show that the algorithms in this paper reduce the recognition time of single traffic sign images by 40.6%, 38.5%, 36.5%, and 33.2%, respectively, compared with the detection recognition algorithm with improved LeNet-5 model on the four different datasets; 23.1%, 18.8%, 24.5%, and 21.0%, respectively, compared with the sign classification algorithm; and 40.3%, 39.3%, and 39.5%, respectively, compared with the AlexNet algorithm 40.3%, 39.2%, 36.6%, 32.7%, respectively; and 43.9%, 39.9%, 35.4%, 32.8%, respectively, compared with the improved deep learning algorithm. Thus, the proposed algorithm significantly outperforms other algorithms in processing time for single traffic sign image recognition on different datasets.

Table 3. Comparison of processing time for single image of different sizes and types of data sets

Algorithm name	CCTSDB-400 sheets Test set	CCTSDB-600 sheets Test set	CCTSDB-700 sheets Test set	CCTSDB-800 sheets Test set
Improved detection recognition algorithm for LeNet-5 model	0.0559 s	0.0596 s	0.0622 s	0.0657 s
Shape classification algorithm based on SVM classifier	0.0432 s	0.0451 s	0.0523 s	0.0556 s
Improving AlexNet algorithm	0.0556 s	0.0602 s	0.0623 s	0.0652 s
Improving deep learning algorithms based on SENet and ResNet models	0.0592 s	0.0609 s	0.0611 s	0.0653 s
Algorithm of this paper	0.0332 s	0.0366 s	0.0395 s	0.0439 s

Comparison of the Ablation of the Algorithms in This Paper on the Evaluation Metrics

Based on the CCTSDB-400 dataset, the study tested using the spatio-temporal component map convolution network and the improved RGB saliency detection spatio-temporal component map convolution network and the improved residual module spatio-temporal component map convolution network, respectively, after preprocessing. The improved RGB saliency detection spatio-temporal component graph convolution network = RGB image saliency detection based on octave convolution + spatio-temporal component graph convolution network, and the improved residual module spatio-temporal component graph convolution network = spatio-temporal component graph convolution network + octave convolution residual module.

As can be seen from Table 4, the algorithm in this paper has the best results in five evaluation metrics after IPT preprocessing of traffic sign images. In the F-measure evaluation index, it improves 19.2%, 15.7%, and 13.7%, respectively, compared with other combination modules; in the Precision evaluation index, it improves 12.5%, 7.3%, and 8.6%, respectively, compared with other combination modules. The results show that the algorithm of this paper takes into account both the recognition accuracy of traffic signs and the recognition speed of vehicles in the driving process.

Table 4. Comparison of evaluation indicators of different module combinations

Module combinations	MAE	RMSE	MAPE	F-measure	Precision
Spatio-temporal component map convolutional network	6.953	9.336	10.1%	78.5%	85.1%
Improved RGB saliency detection spatio-temporal component map convolutional network	6.556	8.390	9.9%	80.9%	89.2%
Improved residual module spatio-temporal component map convolutional network	6.356	8.659	9.4%	82.3%	88.1%
Algorithm of this paper	6.003	7.806	8.9%	93.6%	95.7%

5 Conclusion

A traffic sign image segmentation algorithm based on an improved spatio-temporal map convolutional network is proposed to improve the computational speed and segmentation accuracy of traffic sign image segmentation under extreme weather conditions in multiple scenes. Experimental validation is conducted on three publicly available traffic datasets, and the following conclusions are obtained:

1. In terms of improving the accuracy and computing efficiency, the octave convolution replaces the ordinary convolution of RGB image saliency detection module and octave convolution residual module to greatly reduce the computing burden caused by the complex environment, which realize the lightweight network architecture design, and improves the prediction speed.
2. In the area of spatio-temporal component map convolutional network feature extraction, the application of spatio-temporal component map convolution in traffic was initially proposed in traffic flow prediction, and the study was applied cross-domain in traffic sign recognition, and the common full convolutional module in the original algorithm was replaced with SETR algorithm to improve the computing efficiency significantly.

However, there is still room for further improvement of the algorithm regarding traffic sign clarity enhancement. The next research work will optimize the sign details to achieve faster and more accurate traffic sign image segmentation while improving the overall computing efficiency of the algorithm and reducing the complexity and computational power of the algorithm.

References

1. Wang, K., Zhao, Y., Xing, X.: Deep learning in driverless vehicles. CAAI Trans. Intell. Syst. **13**(1), 55–69 (2018)
2. Chen, F., Liu, Y., Li, S.: Survey of traffic sign detection and recognition methods in complex environment. Comput. Eng. Sci. **57**(16), 65–73 (2021)

3. Deng, X., Zhang, Y., Yang, Y.: A shape recognition algorithm for traffic sign identification. Comput. Eng. Sci. **43**(02), 322–328 (2021)
4. Xu, J., Wang, L.: Traffic sign recognition method based on AlexNet network. Radio-Engineering 1–10 (2022). http://kns.cnki.net/kcms/detail/13.1097.TN.20220111.1637.006.html
5. He, R.-B., Di, L., Liang, J.: An improved deep learning algorithm for road traffic sign recognition. CAAI Trans. Intell. Syst. **15**(6), 1121–1130 (2020)
6. Dewi, C., Chen, R.-C., Tai, S.-K.: Evaluation of robust spatial pyramid pooling based on convolutional neural network for traffic sign recognition system (2020)
7. Cao, J., Song, C., Peng, S., et al.: Improved traffic sign detection and recognition algorithm for intelligent vehicles (2019)
8. Yazdan, R., Varshosaz, M., Pirasteh, S., Remondino, F.: Using geometric constraints to improve performance of image classifiers for automatic segmentation of traffic signs. https://doi.org/10.1139/geomat-2020-0010
9. Di, L., He, R., Liang, J.: Road traffic identification based on probability clustering and convolutional neural network. J. Nanjing Univ. (Nat. Sci.) **55**(02), 238–250 (2019)
10. Jie-Wei, Li-Wei-Sang, Li-Wei: Traffic signs real-time classification and recognition of traffic signs based on multi-feature fusion. Mod. Electron. Technol. **42**(11), 50–53+58 (2019)
11. Mannan, A., Javed, K., Rehman, A.U., et al.: Optimized segmentation and hybrid multiscale feature extraction for traffic sign detection and recognition. J. Intell. Fuzzy Syst. 1–16 (2018)
12. Handoko, H., Pratama, J.H., Yohanes, B.W.: Traffic sign detection optimization using color and shape segmentation as pre-processing system. TELKOMNIKA (Telecommunication Computing Electronics and Control) **19**(1) (2021)
13. James, D.: Computer vision based traffic sign sensing for smart transport. J. Innov. Image Process. **1**(1), 11–19 (2019)
14. Liu, W., Lu, X.: Research progress of transformer based on computer vision. Comput. Eng. Sci. 1–17 (2022). http://kns.cnki.net/kcms/detail/11.2127.tp.20211129.1135.004.html
15. Chen, Q., Hu, Q., Li, J.: Image deblurring method based on dual task convolution neural network. Chin. J. Liquid Cryst. Displays **36**(11), 1486–1496 (2021)
16. Wang, Q.R., Wei, Y.M., Zhu, C.F., et al.: Research on traffic accident risk prediction based on spatio-temporal convolution. Comput. Eng. https://doi.org/10.19678/j.issn.1000-3428.0062961
17. Yao, R., Xia, S.X., Zhou, Y., et al.: Spatio-temporal graph convolutional networks with attention mechanism for video target segmentation. Chin. J. Graph. **26**(10), 2376–2387 (2021)
18. Jiang, T.T., Liu, Y., Ma, X., et al.: Multi-branch collaborative RGB-T image saliency target detection. J. Image Graph. **26**(10), 2388–2399 (2021)
19. Dai, J., Cao, Y., Shen, Q., et al.: Traffic flow prediction based on multi-spatial-temporal graph convolutional network. Appl. Res. Comput. 1–6 (2022). https://doi.org/10.19734/j.issn.1001-3695.2021.08.0361
20. Yu, G., Liu, Z., Zhao, P.: A single image collaborative saliency detection method. Appl. Res. Comput. **37**(S2), 308–310 (2020)
21. Lin, H., Li, J., Liang, D., et al.: Complete detection method of salient objects. J. Chin. Comput. Syst. **37**(09), 2079–2083 (2016)

Research on Sheep Counting Algorithm Under Surveillance Video

Yingnan Wang[1,2,3] and Meili Wang[1,2,3]($\boxtimes$)

[1] College of Information Engineering, Northwest A&F University,
Yangling 712100, China
`wml@nwsuaf.edu.cn`
[2] Key Laboratory of Agricultural Internet of Things, Ministry of Agriculture,
Yangling 712100, China
[3] Shaanxi Key Laboratory of Agricultural Information Perception and Intelligent
Service, Yangling 712100, China

Abstract. Most sheep farms are established in places far away from urban areas, which inconveniences centralized management, and inventory information is an essential task for centralized management. Still, the traditional way of inventory management mainly uses manual statistics is time-consuming and laborious. To address the above problems, inspired by the crowd density counting model SFANet, we propose a sheep counting algorithm under surveillance video based on VGG16 as the frontend feature extractor and the backend as a dual-path multiscale fusion network for generating density maps and attention maps. Attaching ASPP module and CAN module to the back-end density map path not only extracts the multi-scale features of sheep in the image, but also handles their multi-scale variations based on contextual information. By adding the channel attention module SE to the attention map path, the SE channel attention mechanism enhances the channel where the sheep is located according to the importance of each feature channel, thereby making the network more focused on the target. The results show that the network has a mean absolute error of 1.28 and a mean square error of 1.70 under the outdoor sheep farm dataset, while the indoor environment is more complex with a mean absolute error of 5.82 and a mean square error of 7.36. The experimental results show that is helpful for sheep counting under surveillance video.

Keywords: Convolutional neural networks · Sheep counting · Surveillance videos

1 Introduction

With the advancement of information technology, China's sheep farming industry is developing from free-range to large-scale, intensive, and refined style. By

Supported by Shaanxi Province Key R&D Program (2022QFY11-03).

combining modern intelligent technologies such as the Internet of Things, big data, and computer vision with sheep farming, the development of sheep farming can be achieved with higher efficiency and profitability, i.e. smart sheep farm. Smart sheep farms can realize intelligent and precise management, traceability of meat products, and timely monitoring of epidemics [1]. In refined management, sheep counting is an essential task, and its traditional method mainly relies on manual counting, which is time-consuming and labor-intensive. Efficient counting methods help reduce the waste of managers' workforce and better realize the intelligence of sheep farms.

Traditional sheep counting is done automatically by embedded devices in fixed places. Zhang et al. [2] designed a smart sheep counting device based on radio frequency identification (RFID) technology. This intelligent device was used by marking a tag on each sheep's ear and then counting by RFID module. The accuracy of the counting was guaranteed, but using an embedded ear tag would cause damage to the animal's ear, easily lead to wound, and be very inconvenient to use. With the remarkable advantages of convolutional neural network in the field of computer vision, it provides a variety of technical methods for sheep counting. Tian Lei [3] detected and counted the sheep using the YOLOv3 target detection algorithm, and the counting result was poor due to the serious mutual occlusion between sheep. Li et al. [4] Combining YOLOv3 target detection algorithm and Deep SORT tracking algorithm, a double line counting method is designed to realize the automatic counting of grassland sheep. However, the method was harsh for counting sites and required that the direction of the sheep movement was consistent, which was not suitable for indoor free-ranging sheep. Sarwar F et al. [5] used the RCNN network to detect and count sheep in paddocks using UAV video as the study object. The sheep scale variation in this method was slight. However, its detection became poor for the case of considerable sheep scale variation under surveillance video. Du et al. [6] proposed a VDNet network combining visual geometry group (VGG-16) and dilated convolution (DC) to implement the sheep counting method. The network expanded the perceptual field using dilated convolution. It could generate density maps for high-density images, but because the background of their dataset was monotonous, it is prone to the problem of inability to distinguish the background from the target if the targeting dataset has a complex background.

Due to the further growth of convolutional neural networks, more researchers started to propose crowd-counting methods in high-density scenes. Zhang et al. [7] used the three-column convolution with multiscale convolution kernels to cope with the variation of density differences. Li et al. [8] used a CSRNet network that removed the multi-column CNN and used a single-column dilation convolution structure that introduced dilation convolution when dealing with multi-scale problems. Zhu et al. [9] proposed a SFANet network based on dual-path multi-scale fusion decoder decoding for crowd counting. Currently, convolutional neural networks have made great breakthroughs in high-density crowd counting tasks. However, there is no relevant research on sheep counting tasks under sheep farm surveillance videos. Therefore, the article puts forward a method for sheep counting under surveillance video. Our method is based on the SFANet network, and the front-end feature extraction still uses the first 13 layers of VGG16-bn [10]. In the density map path of the dual-path multiscale fusion network, adding ASPP [11] module and CAN [12] module can abstract the multiscale features of sheep in the image and also deal with the problem of multiscale changes according to contextual information. Adding the SE [14] module to the attention map path can better distinguish the sheep in the image from the background. The model not only adapts to the complex environment of sheep farms to achieve accurate counting results but also meets the remote monitoring needs of managers. After the experiments, we have shown that our model is superior to classical approaches in terms of the accuracy of sheep counting.

2   Materials and Methods

2.1   Data Acquisition

Since most of the current sheep datasets are constructed with UAVs as the equipment carrier, this paper will build a sheep counting dataset based on surveillance cameras. The dataset came from a pilot sheep farm in Weinan City, Fuping County. Since the sheep in this experimental base are captive, their range of activities is limited, so there is no need to monitor them in real-time under the surveillance video. The number of sheep in the flock for the entire period can be replaced by an image of a specific video frame. Indoor and outdoor activity areas were selected as the monitoring sites. The equipment used was a Hikvision 4 million UHD, 4mm focal length surveillance camera, with video recorded remotely via PC as the data source. The dataset we produced included 656 images of sheep, half of which are indoor activity areas of sheep farms and half of which are outdoors. The total of sheep in every image is between 10–70. A randomized algorithm was used for two places to use 216 of the 328 images as the training set.

Fig. 1. Image annotation

Pre-processing of the dataset: Firstly, we manually screened the videos recorded by surveillance cameras and then frame extracted the videos to get RGB images of the sheep farm environment. Then the images of indoor and outdoor areas were divided according to the crowd counting dataset Shanghai-Tect A/B [7]. The images suitable for labelling were manually selected and were batch named with python scripts, finally processed images were divided into the testing set and training set. The MATLAB tool was used for manual labelling of this dataset. Due to the "flock effect," some of the sheep's heads were not visible most of the time, so the center of the sheep was chosen to be labelled. As shown in Fig. 1, the central part of the sheep's body with a blue "*" sign indicated that it had been manually labelled.

2.2   Density Map Ground Truth

Considering the small focal length of the surveillance camera and the wide monitoring range will lead to the problem of perspective aberration of the image, i.e., the sheep in the distance appears smaller than the sheep in the immediate area of the same size. Therefore, it is more suitable to use the method of adaptive Gaussian filter to generate the sheep density map [7].

The process of generating the density map from the annotation file is as follows: First, build a matrix with the same size as the original image, set all to 0, and set the corresponding position of each marker head to 1, so as to obtain a matrix with only 0 and 1 [18]. Finally, the 0 and 1 matrices are convolved with Gaussian functions to obtain continuous density maps. The specific equation is as follows.

$$D(x^s) = \sum_{i=1}^{M} \delta \left(x^s - x_i^s \right) \tag{1}$$

where x_i^s marks the position of the point for each sheep and $D(x^s)$ is an image label with M sheep. Here x^s represents a two-dimensional coordinate, so $D(x^s)$ is a matrix of 0 and 1.

The density map in $D(x^s)$ is discrete and smoothed with a Gaussian function $G(x)$ to convert it into a continuous function. The actual image density map $F(x^s)$ can be expressed as follows.

$$\begin{cases} F(x^s) = D(x^s) \star G_{\sigma_i}(x) \\ \sigma_i = \beta \bar{d}_i \end{cases} \tag{2}$$

where $\bar{d}_i$ is the average distance between the marker point x_i^s and the nearest K sheep. β is the empirical parameter and takes 0.3 in the paper.

The sheep density map is labelled with the center of the sheep's body as its position. The distribution of sheep positions is shown in the form of notable highlights, and the integral value of the density map represents the total of sheep in this image. Figure 2 shows the sheep density map after the adaptive Gaussian kernel processing. The sum of probabilities for each sheep body part region is 1. After obtaining the complete sheep image density map, the number of sheep is obtained by integrating and summing them [13].

2.3   Proposed Method

We propose a dense sheep counting model based on the improved SFANet [9] model of dense crowd counting. This network adds ASPP [11] module and CAN

Fig. 2. Flock density map generation

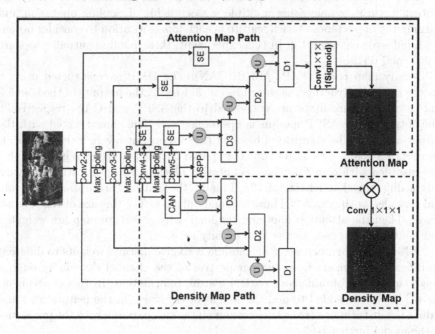

Fig. 3. The framework of our method

[12] module for density map path and SE [14] module for attention map path based on SFANet network structure. The specific network structure diagram is shown in Fig. 3.

Feature map extractor (VGG16-bn) [10]: The front-end extractor selects the first pretrained 13-layer network of VGG16-bn to acquire multi-scale features in the image, adopting all 3*3 filters in the network because it can extract different scales features and semantic information [13]. Four layers: conv2-2, conv3-3, conv4-3 and conv5-3, are connected to the SE module. The output of the conv5-3 and conv4-3 layers are used as the input of the ASPP and CAN modules, respectively. The ASPP module is composed of an atrous convolution, using multiple parallel expansion convolution layers with various sampling rates. The problem of loss of image detail information due to the small size of the output image of the conv5-3 feature map is avoided by the multiscale perception module ASPP. The CAN module generates scale-aware contextual features using multiple receiver domains of the averaging pooling operation, preserving the scale features as well as the contextual information of the image [13].

ASPP (Atrous Spatial Pyramid Pooling) [11]: ASPP module introduces cavity convolution with distinct atrous rates, thus aggregating multiscale context information to enhance the capacity of the model to recognize the same object of different sizes. In this experiment, the atrous rates are 1, 12, 24, and 36, and the convolution kernel size is 3*3.

CAN (Context-aware module): CAN module can generate scale-perceiving context features by averaging multiple sensory fields of pooling operation and learning the importance of each feature in each image location to consider potential rapid scale changes [13]. In this experiment, these pooling output scales are 1, 2, 3, and 6 [12].

Density map path (DMP+ASPP+CAN): The DMP is constructed using a feature pyramid structure, as shown in Fig. 3. Firstly, the outputs of the conv5-3 and conv4-3 feature maps are connected to the ASPP and CAN, respectively. The output of the ASPP module is upsampled and then concatenated with the output of CAN. The connection block of the two output layers includes concat, $conv1 \times 1 \times 256$ and $conv3 \times 3 \times 256$. The spliced output feature map is upsampled and cascaded with conv3-3, whose connection block has a similar structure with only a different channel size of 128. Finally, the features are upsampled again and cascaded with conv2-2. Thus, the final output is $1/2$ the size of the original image. Then the attention map and the final density feature map are weighted and multiplied to generate the final density map.

SENet [14]: The purpose of SE module is to give distinct weights to different positions of the image from the perspective of the channel domain through a weight matrix to obtain more critical feature information. In this experiment, the SE module is added to add the weight of the sheep in the feature map and reduce the disturbance of the background on the sheep, to improve the properties of the model further [16].

Attention map path (AMP+SE): AMP is also constructed using a feature pyramid structure. Four source layers are applied as the input of the SE module, and the multi-scale output features are integrated into the attention map. The SE module enhances the channels containing sheep in the image by modeling the significance of every feature channel and suppressing the interference of the sheep field background, thus improving the sheep odds of every pixel point in the density map.

3   Experiments Setup

3.1   Experiment Preparation

The server operating system used for this experiment is Ubuntu 18.04. The central processor is NVIDIA TITAN RTX with Cuda 11.3. MATLAB 2018b is used as the flock dataset annotation tool. The specific software and hardware configurations are shown in Table 1.

Table 1. Experimental software and hardware configuration

Software and hardware types	Parameter values
Operating system	Ubuntu 18.04(64 bit)
Deep learning framework	PyTorch 1.10.0
Development languages	Python 3.7.10
Central processing unit	Intel Xeon Platinum 8160T @2.1 GHz
Memory	128 G
Video card	NVIDIA TITAN RTX 24 G

3.2 Evaluation Metrics

The current evaluations for dense counting models are done by computing the mean absolute error (MAE) and the mean square error (MSE), so this paper also uses them to appraise the model accuracy [9,17].

$$MAE = \frac{1}{N}\sum_{i=1}^{N}\left|C_i^P - C_i^G\right| \tag{3}$$

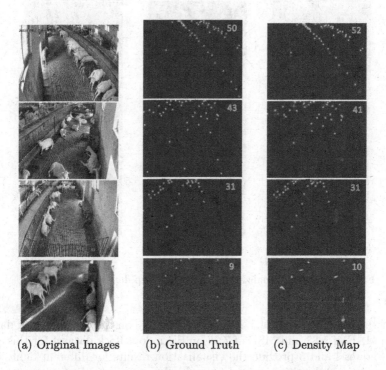

(a) Original Images   (b) Ground Truth   (c) Density Map

Fig. 4. Indoor: visualization results of sheep data set density map

$$MSE = \sqrt{\frac{1}{N}\sum_{i=1}^{N}\left|C_i^P - C_i^G\right|^2} \tag{4}$$

where N is the sum of test images. C_i^P and C_i^G are the predicted and true number of flock image, respectively. The following equation C_i^P is the calculation process:

$$C_i^P = \sum_{w=1}^{W}\sum_{d=1}^{D} Z_{w,d} \tag{5}$$

where W and D denote the length and width of the density map, and $Z_{w,d}$ denotes the pixel points at the density map.

4   Results and Analysis

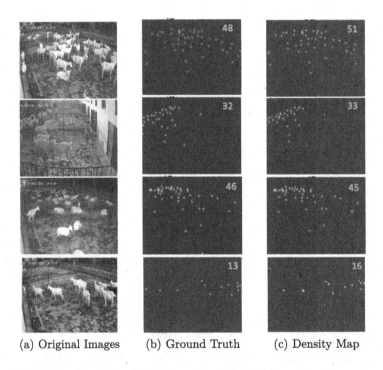

(a) Original Images   (b) Ground Truth   (c) Density Map

Fig. 5. Outdoor: visualization results of sheep data set density map

Our method had an MAE of 1.28 and MSE of 1.70 on the outdoor flock dataset and an MAE of 5.82 and MSE of 7.36 on the indoor flock dataset for this sheep farm. Figures 4 and 5 provide the visualization results for different light conditions, angles and densities in outdoor and indoor areas. In the figures, (a) is the primitive picture, (b) is the image of the sheep density map generated from the actual location of the sheep in the picture, and (c) is the density map obtained after the prediction of the model in this paper.

Since the experimental data contains images at different angles, illuminations, and densities, it can be seen from the partial results test in Fig. 4 and Fig. 5 that the prediction error is still small, whether it is in the case of a large density or a small number of them. In the case of insufficient light or overexposed, our method can predict its quantity very well. It shows that the prediction outcome of our proposed model match well with the actual images and has good robustness to the backgrounds such as light and occlusion in the images, and can reflect the distribution of sheep in the input images accurately.

In Table 2, we contrasted several ways often used as the standard for crowd counting algorithms, this method significantly improved. Compared with the deep convolutional neural network (CAN), the MAE of this method in part A of shanghai Tech dataset decreases by 2.9%, and the MAE in part B of the dataset decreases by 6.4%.

Table 2. Ablation experiments on the ShanghaiTechA/B dataset [7]

Methods	Part_A		Part_B	
	MAE	MSE	MAE	MSE
MCNN [7]	110.2	173.2	26.4	41.3
CSRNet [8]	68.2	115.0	10.6	16.0
SANet [15]	67.0	104.5	8.4	13.6
CAN [12]	62.3	100.0	7.8	12.2
Ours	**60.5**	105.9	**7.3**	13.1

Table 3 shows the consequence of comparing the method of this paper in the sheep dataset with the base model SFANet, and also the base model with the addition of different modules. Compared with the base model SFANet, the method of this paper can reduce MAE and MSE by 16.9% and 15% on the outdoor sheep dataset, but on the indoor sheep dataset, the method of this paper has the optimal result without the SE module, which can reduce MAE and MSE by 4.6% and 2.2%. It shows in Table 2 that this method is effective on the usual crowd counting ShanghaiTech dataset, while the reasons for the poor performance on the sheep dataset in the indoor area are: first, there could be errors in manual labelling; second, the overexposure of the image could be confused with the color of the sheep.

Table 3. Experimental comparison on the sheep dataset

	Outside		Inside	
Methods	MAE	MSE	MAE	MSE
SFANet	1.54	2.00	4.54	5.87
SFANet (CAN)	1.38	1.98	4.44	5.76
SFANet (ASPP)	1.37	1.84	4.51	5.81
SFANet (CAN+ASPP)	1.43	1.96	**4.33**	**5.74**
SFANet (CAN+ASPP+SE)	**1.28**	**1.70**	5.82	7.36

5　Conclusion

On the basic crowd counting model SFANet, we propose a sheep counting method under surveillance video. The frontend network is the first 13 layers of VGG16, and the backend is a dual-path multiscale fusion network, which finally generates a density map and an attention map. The ASPP module and CAN module are attached to the density map path, which can extract the multi-scale features of sheep in the images but also deal with their multiscale variations according to the contextual combination. Connecting the attention map path to the SE module suppresses the interference of the sheep field background. It does not need to detect sheep in images as traditional detection-based counting methods, and it solves the problem of decreasing counting accuracy in occlusion cases. After experimental verification, the average absolute error of the counting algorithm in this paper is 1.28, and the mean square error is 1.70 in the outdoor activity area under the monotonous background of the sheep farm. The MAE is 5.82, and the MSE is 7.36 in the indoor activity area under the complex background. The experiments verify that the way in our paper has better performance for flock counting. In the following research step, a more efficient attention module will be designed to solve the false detection caused by light exposure.

Acknowledgments. This work was supported by Key Research and Development Projects of Shaanxi Province (grant 2022QFY11-03) and Science and Technology Think Tank Young Talent Program (grant 20220615ZZ07110314)

References

1. Yao, Z., Tan, H., Tian, F., Zhou, Y., Zhang, C.: Research progress of computer vision technology in intelligent sheep farm. China Feed **07**, 7–12 (2021)
2. Zhang, L., Jing, X., Tian, Z., Meng, L., Zhao, Y.: Research and implementation of intelligent counting sheep system in pastoral areas. Commun. Power Technol. **04**, 165–166 (2017)
3. Tian, L.: Design of flock size detection system . Inner Mongolia University, Huhehaote (2019)

4. Li, Q., Shang, J., Li, B.: An automatic counting method for grassland sheep based on head image features. China Testing (11), 20–24 (2020)
5. Sarwar, F., Griffin, A., Periasamy, P., Portas, K., Law, J.: Detecting and counting sheep with a convolutional neural network. In: 2018 15th IEEE International Conference on Advanced Video and Signal Based Surveillance (AVSS), pp. 1–6 (2018)
6. Yongxing, D., Miao, X., Qin, L., Li, B.: VDNet convolutional neural network based flock counting. Laser Technol. **05**, 675–680 (2021)
7. Zhang, Y., Zhou, D., Chen, S., Gao, S., Ma, Y.: Single-image crowd counting via multi-column convolutional neural network. In: Proceedings of the IEEE Conference on Computer Vision and Pattern Recognition, pp. 589–597 (2016)
8. Li, Y., Zhang, X., Chen, D.: CSRNet: dilated convolutional neural networks for understanding the highly congested scenes. In: Proceedings of the IEEE Conference on Computer Vision and Pattern Recognition, pp. 1091–1100 (2018)
9. Zhu, L., Zhao, Z., Lu, C., Lin, Y., Yao, P., Tangren, Y.: Dual path multi-scale fusion networks with attention for crowd counting. arXiv: Computer. Vision and Pattern Recognition (2019)
10. Simonyan, K., Zisserman, A.: Very deep convolutional networks for large-scale image recognition. arXiv preprint arXiv: 1409.1556 (2015)
11. Chen, L.-C., Papandreou, G., Kokkinos, I., Murphy, K., Yuille, A.L.: DeepLab: semantic image segmentation with deep convolutional nets, Atrous convolution, and fully connected CRFs. IEEE Trans. Pattern Anal. Mach. Intell. **40**(4), 834–848 (2016)
12. Liu, W., Salzmann, M., Fua, P.: Context-aware crowd counting computer vision and pattern recognition. In: Proceedings of the IEEE/CVF Conference on Computer Vision and Pattern Recognition, pp. 5099–5108 (2018)
13. Thanasutives, P., Fukui, K.I., Numao, M., Kijsirikul, B.: Encoder-decoder based convolutional neural networks with multi-scale-aware modules for crowd counting. In: 2020 25th International Conference on Pattern Recognition, pp. 2382–2389 (2021)
14. Jie, H., Li, S., Albanie, S., Sun, G., Enhua, W.: Squeeze-and-excitation networks computer vision and pattern recognition. IEEE, Piscataway (2018)
15. Cao, X., Wang, Z., Zhao, Y., Su, F.: Scale aggregation network for accurate and efficient crowd counting European conference on computer vision, pp. 734–750 (2018)
16. Wang, Z., Xie, K., Zhang, X., Chen, H., Wen, C., He, J.: Small-object detection based on YOLO and dense block via image super-resolution. IEEE Access **9**, 56416–56429 (2021)
17. Xingjiao, W., Kong, S., Zheng, Y., Ye, H., Yang, J., He, L.: Feature channel enhancement for crowd counting. IET Image Proc. **14**(11), 2376–2382 (2020)
18. Yu, J., Jia, R., Li, Y., Sun, H.: Automatic fish counting via a multi-scale dense residual network. Multimedia Tools Appl. **81**(12), 17223–17243 (2022). https://doi.org/10.1007/s11042-022-12672-y

A Blockchain-Based Distributed Machine Learning (BDML) Approach for Resource Allocation in Vehicular Ad-Hoc Networks

Dajun Zhang[1]([✉])[ID], Wei Shi[1][ID], and Ruizhe Yang[2][ID]

[1] Carleton University, 1125 Colonel By Drive, Ottawa, Canada
dajunzhang9038@gmail.com
[2] Beijing Laboratory of Advanced Information Networks,
Beijing University of Technology, Beijing, China

Abstract. Recently, effective allocation of VANET resources is a key factor in promoting the development of VANETs. Due to high bandwidth costs, poor time efficiency, and a high risk of privacy leakage, the use of traditional centralized data centers to analyze massive data has proven to be a difficult task. These challenges have prompted a revolutionary change in VANET architectures to scatter computations from a centralized data center to distributed network edges. Distributed VANET configurations leverage the computing power of network edges by using a large number of mobile devices which frequently exchange data with the edge of the network or among themselves. However, the heterogeneity and distrust of the distributed edge hinder the efficient, reliable, and secure allocation of VANET resources. In this paper, we express the allocation strategy for both computing and network resources as a joint optimization problem. We use a local deep reinforcement learning with a prioritized experience replay mechanism on edge nodes and use the blockchain for sharing the optimal learning results to optimize the overall resource allocation problem. Simulation results show that our proposed scheme is superior to a current machine learning approach.

Keywords: Vehicular ad hoc networks · Blockchain · Deep reinforcement learning

1 Introduction

Recently, vehicular ad-hoc networks (VANETs) have aroused great interest. In the VANET environment, the limitations of network resources prevent the improvement of edge computing. Currently, due to privacy concerns and the frequent data loss during high-speed streaming data transmission, data shared between VANET users are limited to only emergency situations. Furthermore, the limitation of edge computing power brings new challenges and difficulties to resource allocation in VANETs [1]. For example, the uncertainty of vehicle movement may lead to competition between VANET nodes (i.e. the computing

C. Yu et al. (Eds.): GPC 2022, LNCS 13744, pp. 110–121, 2023.
https://doi.org/10.1007/978-3-031-26118-3_8

unit on board of a vehicle) for the computing power of edge nodes (i.e. RSUs) resulting in the insufficient allocation of computing capability. Considering the sparse deployment of RSUs, when the number of vehicles within the coverage of a RSU increases rapidly, the resource spectrum of the RSU cannot meet the needs of all vehicles. Therefore, fair and effective allocation of computing resources and network resources to VANET nodes has become an urgent problem to be solved.

At this stage, some scholars have put forward resource allocation mechanisms in VANETs. The research [2] proposed the allocation of computing resources for video processing, aiming at improving the quality of service for users. The authors of [3] proposed a VANET framework for the dynamic adjustment of network, cache, and computing resources, aiming to improve the joint resource allocation problem of existing VANETs. The authors of [4] designed a spectrum-sharing scheme in order to solve the data conflict between the cellular network of vehicles and users. For the occupation of unlicensed channels, the two can fairly compete for the right to use the channel. In [5], the author proposes a new decomposition algorithm that utilizes integer linear programming to fully allocate computing resources. It provides ideas for the future software-defined VANET data-sharing architecture. In order to ensure the correct deployment of the blockchain in the distributed software-defined Internet of Vehicles, the authors of [6] propose a distributed software-defined VANET architecture based on the blockchain to ensure the utility of resource allocation.

In this article, we propose an integrated resource allocation scheme for the VANETs based on the blockchain and Deep Reinforcement Learning (DRL). The contributions of this article are summarized as follows:

- We propose a new resource (network resource and computing capability) allocation scheme based on blockchain. Each VANET node trains a neural network locally through the DRL algorithm to obtain resource allocation strategies and shares the best local learning results through distributed edge nodes. Here, we aim to use the communication utility to characterize the quality of the communication link between the node and the RSU, so as to allocate the best RSU to neighboring nodes for local training. Meanwhile, the computing utility is used to characterize the utilization of computing resources. The VANET node trains the joint resource utility locally within a fixed time, and then uploads the result to the blockchain system, aiming to improve the joint resource utility through the parameter sharing of the blockchain system. The core advantage of using the blockchain for sharing is that the blockchain is a completely decentralized system, and the local training results are stored in the transaction. The non-tamperable feature of the blockchain ensures the safety of local training results. The proposed scheme lays a new foundation for intelligence sharing between edges.
- We make improvements to the traditional DRL algorithm in local training. Aiming at solving the problem that the traditional DRLs experience in inefficient random sampling using replay strategy, we have introduced a prioritized experience replay strategy to the local training. We define the priority of the

samples and prioritize the samples in the experience replay pool to improve the efficiency of local training.
- Finally, we have adopted a Redundant Byzantine Fault-Tolerant (RBFT) [7] mechanism. Since a permissioned-blockchain system is used, RBFT has better fault tolerance and better overall performance than the Practical Byzantine Fault-Tolerant (PBFT) consensus mechanism.

The rest of the paper is organized as follows: the system model is described in Sect. 2. Section 3 illustrate the problem formulation of the local training. In Sect. 4, we introduce the local training process and blockchain sharing mechanism. In Sect. 5, we present the experiments that have been conducted in this work. Finally, conclusions are described in Sect. 6.

2   System Description

In this section, we illustrate the system model in our proposed framework. We discretize the time T into a number of time slot $\mathcal{T} = \{1, 2, ..., t, ..., T\}$.

2.1   Communication Model

Figure 1 shows that we use collaborative edge computing to balance the loss of high-speed streaming data transmission and the overhead on transmission. The proposed framework has $\mathcal{C} = \{R_1, R_2, ..., R_c, ..., R_C\}$ Road Side Units (RSUs) and $\mathcal{E} = \{M_1, M_2, ..., M_e, ..., M_E\}$ mobile edge servers. In our proposed architecture, U vehicles around multiple RSUs and edge servers communicate with each other, denoted by $\mathcal{U} = \{V_1, V_2, ..., V_u, ..., V_U\}$. In this paper, we implement blockchain sharing of local training results (communication and computation utilization) of multiple vehicles.

The communication model is defined as finite-state Markov channels (FSMCs) [8]. The channel state $\delta^{V_u, R_c}(t)$ between V_u and R_c depends on the received signal-to-noise ratio (SNR), which is divided into K levels that is defined as $\mathcal{F} = \{F_0, F_1, ..., F_{k-1}, ..., F_K\}$.

Since the communication channel is a time-varying channel, we can set the variation of SNR follow a Markov decision process. Let $p_{k,k'} = Pr\{\delta^{V_u, R_c}(t+1) = F_{k'} | \delta^{V_u, R_c}(t)\}$ be the channel state transition probability, where $k, k' \in \mathcal{K}$. Hence, $\mathcal{P} = [p_{k,k'}]_{K \times K}$ is the channel state transition matrix of the $K \times K$ dimension.

We define the available bandwidth allocated to V_u as $H_{R_c}^{V_u}$. According to Shannon's theorem, the maximum communication rate $r_{R_c}^{V_u}(t)$ is:

$$r_{R_c}^{V_u}(t) = H_{R_c}^{V_u}(t) \log_2(1 + \delta^{V_u, R_c}(t)) \tag{1}$$

Therefore, we define the communication resource utilization of V_u as the ratio of $r_{R_c}^{V_u}(t)$ to the available bandwidth $H_{R_c}^{V_u}$ allocated to V_c. Hence, we have:

$$D_{V_u}^{Comm}(t) = \sum_{R_c=1}^{R_C} a_{V_u}^{R_c}(t) \frac{\sigma_{V_u} r_{R_c}^{V_u}(t)}{\theta_{R_c} H_{R_c}^{V_u}(t)} \tag{2}$$

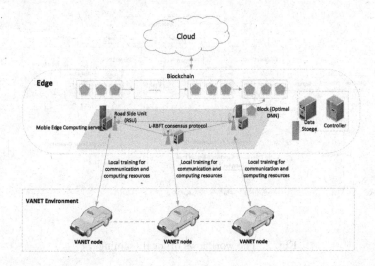

Fig. 1. Blockchain-based resource allocation framework of VANETs.

where σ_{V_u} means the revenue per unit communication rate that V_u can achieve, and θ_{R_c} denotes the unit payment for using networking resources from R_c. We will illustrate the action value $a_{V_u}^{R_c}(t)$ in Sect. 3.

2.2 Computation Model

We use $\epsilon_{V_u}^{M_e}(t)$ to represent the computing capability of each M_e assign to V_u at time slot t, which satisfy the Markov decision process. We discretize random variable $\epsilon_{V_u}^{M_e}(t)$ into Z state, indexed by $\Lambda = \{\Lambda_0, \Lambda_1, ..., \Lambda_{Z-1}\}$. The state transition probability $g_{z'z} = Pr\{\epsilon_{V_u}^{M_e}(t+1) = \Lambda_z | \epsilon_{V_u}^{M_e}(t) = \Lambda_{z'}\}$ determines the state change of $\epsilon_{V_u}^{M_e}(t)$, where $\Lambda_z, \Lambda_{z'} \in \Lambda$.

We usually use time delay to characterize the Quality of Service (QoS) in VANETs. We quantify QoS in terms of queuing delay and propagation delay [9]. We refer the queuing model proposed in [10], we have:

$$T_{qd}^{V_u} = \frac{1}{\epsilon_{V_u}^{M_e}(t) - \lambda_{V_u}(t)} - \frac{1}{\lambda_{V_u}(t)} \cdot \frac{Q_L^{V_u} \rho^{Q_L^{V_u}}}{1 - \rho^{Q_L^{V_u}}} \qquad (3)$$

where workload arrival rate is denoted as $\lambda_{V_u}(t)$, $\rho^{Q_L^{V_u}} = \lambda_{V_u}(t)/\epsilon_{V_u}^{M_e}(t)$ is the utilization, and the maximum queue length is $Q_L^{V_u}$.

The propagation delay T_{pd} can be defined as:

$$T_{pd}^{V_u} = d_{V_u}^{M_e}/\gamma \qquad (4)$$

where γ is propagation speed and $d_{V_u}^{M_e}$ is the physical distance between $V_u(x_u, y_u)$ and $M_e(x_e, y_e)$. Hence, $d_{V_u}^{M_e}$ can be defined as $d_{V_u}^{M_e} = \sqrt{(x_e - x_u)^2 + (y_e - y_u)^2}$.

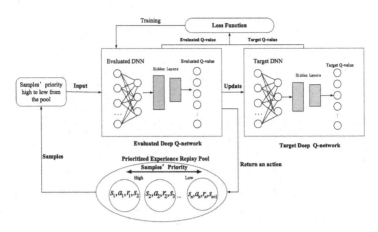

Fig. 2. The workflows of local training.

Finally, we use $T_{V_u} = T_{qd}^{V_u} + T_{pd}^{V_u}$ to characterize the service delay of the overall system.

Therefore, the computing resource utilization $D_{V_u}^{Comp}(t)$ is:

$$D_{V_u}^{Comp}(t) = \sum_{M_e=1}^{M_E} a_{V_u}^{M_e}(t) \frac{\zeta_{V_u}\lambda_{V_u}}{\alpha_{V_u}k_{M_e}o_{V_u} + \beta_{V_u}T_{V_u}} \tag{5}$$

where α_{V_u} and β_{V_u} are the weight ratios of relevant parameters. Here, ζ_{V_u} represents the benefits that V_u can obtain due to the workload of operating unit, k_{M_e} is the operating overhead per unit CPU cycle, and o_{V_u} is the number of CPU cycles required for the computation task. We will illustrate the meaning of action value $a_{V_u}^{M_e}(t)$ in Sect. 3.

3   Problem Formulation

In this section, we discuss the resource allocation problem of VANET node in detail by defining the state space, action space and reward function.

3.1   System State

We set a parameter $s(t), t \in \mathcal{T}$ represent the system state space of proposed framework at time slot t. $s(t)$ includes two components: $\delta^{V_u,R_c}(t)$ and $\epsilon_{V_u}^{M_e}(t)$. Here, we have:

$$s(t) = \begin{bmatrix} \delta^{V_1,R_1}(t) & \cdots & \delta^{V_1,R_c}(t) & \cdots & \delta^{V_1,R_C}(t) \\ \delta^{V_2,R_1}(t) & \cdots & \delta^{V_2,R_c}(t) & \cdots & \delta^{V_2,R_C}(t) \\ \cdots & \cdots & \cdots & \cdots & \cdots \\ \delta^{V_U,R_1}(t) & \cdots & \delta^{V_U,R_c}(t) & \cdots & \delta^{V_U,R_C}(t) \\ \epsilon_{V_1}^{M_1}(t) & \cdots & \epsilon_{V_1}^{M_e}(t) & \cdots & \epsilon_{V_1}^{M_E}(t) \\ \epsilon_{V_2}^{M_1}(t) & \cdots & \epsilon_{V_2}^{M_e}(t) & \cdots & \epsilon_{V_2}^{M_E}(t) \\ \cdots & \cdots & \cdots & \cdots & \cdots \\ \epsilon_{V_U}^{M_1}(t) & \cdots & \epsilon_{V_U}^{M_e}(t) & \cdots & \epsilon_{V_U}^{M_E}(t) \end{bmatrix} \tag{6}$$

3.2 System Action

We set $a(t), t \in \mathcal{T}$ as the action space of the system. The definition of $a(t)$ is:

$$a(t) = \begin{bmatrix} a_{V_1}^{R_1}(t) & \cdots & a_{V_1}^{R_c}(t) & \cdots & a_{V_1}^{R_C}(t) \\ a_{V_2}^{R_1}(t) & \cdots & a_{V_2}^{R_c}(t) & \cdots & a_{V_2}^{R_C}(t) \\ \cdots & \cdots & \cdots & \cdots & \cdots \\ a_{V_U}^{R_1}(t) & \cdots & a_{V_U}^{R_c}(t) & \cdots & a_{V_U}^{R_C}(t) \\ a_{V_1}^{M_1}(t) & \cdots & a_{V_1}^{M_e}(t) & \cdots & a_{V_1}^{M_E}(t) \\ a_{V_2}^{M_1}(t) & \cdots & a_{V_2}^{M_e}(t) & \cdots & a_{V_2}^{M_E}(t) \\ \cdots & \cdots & \cdots & \cdots & \cdots \\ a_{V_U}^{M_1}(t) & \cdots & a_{V_U}^{M_e}(t) & \cdots & a_{V_U}^{M_E}(t) \end{bmatrix} \tag{7}$$

where $a_{V_u}^{R_c}(t)$, and $a_{V_u}^{M_e}(t)$ are:

1) The value of $a_{V_u}^{R_c}(t)$ is 1 or 0. For example, if the action value of R_c is 1 at time t, then RSU and V_c communicate with each other. Otherwise, the action value of R_c is 0. In this paper, we assume that V_u can only communicate with one RSU in a time slot. So we can get $\sum_{R_c=1}^{R_C} a_{V_u}^{R_c}(t) = 1$.

2) The value of $a_{V_u}^{M_e}(t)$ is $\{0,1\}$. For example, if $a_{V_u}^{M_e}(t) = 0$ at time slot t, then M_e does not need to calculate the computing task from V_u. Otherwise, the action value of $a_{V_u}^{M_e}(t)$ is 0. We assume that only one M_e handles the computation task in a time slot. So we can get $\sum_{M_e=1}^{M_E} a_{V_u}^{M_e}(t) = 1$.

3.3 Reward Function

The reward function is accomplished through decision-making in $a(t)$ and $s(t)$. Then, the immediate reward function (utility) of the system is:

$$r_{V_u}(t) = \sum_{V_u=1}^{V_U} (D_{V_u}^{Comm}(t) + D_{V_u}^{Comp}(t)) \tag{8}$$

4 Blockchain-Based Sharing Strategy

In this section, we describe the local training strategy of VANET nodes and the details of the blockchain-based sharing mechanism.

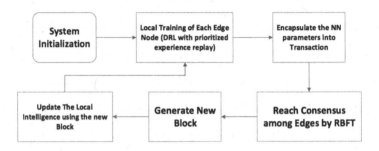

Fig. 3. The sharing mechanism used in BDML

4.1 Local DRL Training with Prioritized Experience Replay of VANET Node

In the deep reinforcement learning experience replay pool, random sampling may repeatedly select redundant samples and reduce training efficiency. Therefore, we have added the Prioritized Experience Replay mechanism (PER) to local learning. The core idea of PER is to prioritize samples. In other words, this method can find samples more efficiently. Figure 2 shows the local training process of the proposed framework. In PER, the probability of each sample (e.g., $(s(t)$, $a(t)$, $r_{V_u}(t)$, $s(t+1))$ being selected is monotonic according to priority. Specifically, the probability of extracting the transition marked j is defined as:

$$P(j) = \frac{p_j^\tau}{\sum_k p_k^\tau} \qquad (9)$$

where p_j^τ is the priority of sample j.

In PER, temporal-difference (TD) errors are often used to prioritized samples [11]. In other words, the premise that a sample has higher priority for the agent to learn is that the TD error is larger. Therefore, we have:

$$\rho_j = R_j + \gamma_j Q_{target}\left(S_j, \arg\max Q(S_j, a)\right) - Q(S_{j-1}, A_{j-1}) \qquad (10)$$

where R_j and $Q(S_j, a)$ are the reward value and Q-value of sample j respectively.

After we determine the TD error j, we need to define importance-sampling weight (IS-weight) as:

$$w_j = \left(\frac{1}{G} \cdot \frac{1}{P(j)}\right)^\kappa \qquad (11)$$

where κ is responsible bias adjustment and G represents the size of experience replay memory.

4.2 Blockchain Sharing Mechanism

In our proposed architecture, RSUs and internal edge computing nodes form the basic structure of the blockchain. The block contains a batch of transactions, and the consensus mechanism is responsible for unified packaging and sequencing. After receiving a block, the blockchain node executes transactions in sequence based on the original account status and reads/writes the status data of the relevant account during this period. The completion of a transaction execution means that the state of the blockchain has undergone a change.

For each transaction, there will be a corresponding transaction receipt or illegal transaction record in the blockchain to indicate the final execution result. If the transaction is a legal transaction, after the execution ends, the result of the transaction execution will be recorded in the transaction receipt. Otherwise, the reason for the error will be recorded in an illegal transaction record. For every illegal transaction, the error information will be encapsulated into an illegal transaction record and stored locally on the node. In addition to the transaction data related to it, the illegal record will also record the specific error reason. For example, if a malicious VANET node sends a false local training result, it will not have sufficient authority to execute the smart contract.

The transaction is initiated by an external user (VANET node). The locally trained VANET node encapsulates the training results (DNN parameters and the training loss) in the transaction and uploads them to the nearby RSU node. When the blockchain node receives the transaction and verifies it first, the node will only process the verified request. The node will do the following transaction verification: (1) Verify the legality of the transaction field, including the transaction format and the legality of the timestamp; (2) Whether the same transaction has been submitted; (3) Verify the transaction signature. After the transaction passes the above verification, a consensus is reached between the blockchain nodes, and the received transaction is broadcast to the consensus nodes of the entire network.

1) *Pre-Prepare*: The Pre-Prepare consensus master node will sequence the transactions within a certain period of time (or a certain number), package them into a block, and then send them to the entire network for consensus.
2) *Preparation*: Prepare all consensus nodes preprocess the block and broadcast the resulting hash.
3) *Submit*: Commit all consensus nodes to write to the block and update the blockchain ledger.

Illegal transactions discovered during the execution process will be stored in the illegal transaction records of the database, and will not be recorded on the blockchain ledger. All legal transactions are stored on the blockchain ledger. When consensus is reached, the node verifies that the transaction and block are correct and valid. The node will automatically execute the smart contract. The node uses its own private key to sign the content of the transaction initiator and the transaction receiver to prevent the content of the transaction from being tampered with. After verifying the signature, MAC, and smart contract, the system sends the newly generated block back to the edge learning node and attaches the block to the blockchain. The edge learning node analyzes the payload information, that is, each edge learning node learns the parameters of

Table 1. Simulation parameters

Simulation parameter	Assigned value
Number of vehicles	8
σ_{V_u}	8 units/Mbps
θ_{R_c}	2 units/Mbps
$H_{R_c}^{V_u}$	4 Mbps
λ_{V_u}	90 Mbps
ζ_{V_u}	3 unit/Mbps
α_{V_u}	0.5
β_{V_u}	0.5
k_{M_e}	2 w/Mcycles
o_{V_u}	50 Mcycles

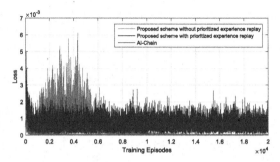

Fig. 4. The convergence performance comparison of three schemes.

the deep neural network from other learning nodes to share the optimal neural network parameters. Figure 3 shows the procedure of the blockchain sharing mechanism.

5   Simulation Results and Discussions

In this section, we verify the superiority of the proposed framework through Tensorflow [12] simulation. Deep Q-learning with a prioritized experience replay mechanism is adopted in the local training process.

5.1   Simulation Setup

Here, we need to point out in particular that the communication utility and computing utility are the key elements that determine the system resource allocation. The communication utility model determines the RSU assigned to a certain VANET node and the channel state of this node. The computing utility model determines the computing capability of the edge computing server

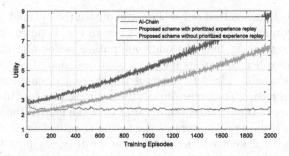

Fig. 5. The utility comparison of three schemes.

allocated to the VANET node for local training. Specifically, RSUs with better channel conditions and higher-power edge computing servers should be used to perform computing tasks. We use these two factors as the local training reward function to characterize the local training effect. Share the parameters of the local training model through the blockchain to enhance the utility of the overall system. Therefore, we treat the problem of joint resource allocation as proof of the effectiveness of our proposed architecture.

In our simulation, we use AI-chain [13] as a comparison scheme to demonstrate the superiority of our proposed scheme. Meanwhile, we have made some adjustments to the existing AI-chain. First, we apply AI-chain to the VANET environment. Secondly, we replaced the neural network for the local training part of AI-chain from the traditional deep neural network to Convolutional Neural Network (CNN). The purpose of this is to meet the comparison requirements of our simulation. The configuration of the AI-chain blockchain adopts a consensus mechanism called learning proof (permissionless), and the local training process follows the traditional deep neural network. The basic idea of AI-chain is to share the optimal neural network parameters among edge nodes through the consensus mechanism of learning proof to meet the needs of joint resource allocation. Therefore, [13] meets the requirements as a comparison scheme. Table 1 shows the simulation parameters in this paper.

5.2 Simulation Results

Figure 4 shows the fitting effect of the convergence curve characterized by the loss function under different schemes. We can see that the curve oscillates significantly at the beginning of the gradient descent. After a training period, the amplitude of the curve oscillations tends to flatten, which means that the optimal policy will be found when the training curve converges. The blockchain shared resource allocation scheme we proposed is different from DRL in that we use multiple edge computing nodes as providers of training results, and solve the joint resource allocation problem of the network by sharing learning results. Through these comparisons, the convergence performance of the blockchain shared resource allocation scheme we proposed has a better convergence effect, which means that our scheme can obtain the optimal strategy faster after a short training, as shown in the red curve in Fig. 4 shown. The reason for the rapid convergence is that the edge computing nodes obtain better DNN parameters from other consensus nodes through the shared parameter mechanism of the blockchain. The advantage of the sharing mechanism is to reduce the waste of computing resources during the training process. In addition, the rapid convergence of the red curve shows that with the help of the sharing mechanism, the local training node iterates the actual Q-value faster, and it is easier to obtain the optimal joint resource allocation strategy, saving a lot of training time. In addition, the prioritized experience replay mechanism is better than the scheme without it. This is because the priority experience replay strategy sets the priority of the system samples, and the system trains according to the priority of the samples. It avoids the drawbacks of repeated training of samples in the experience replay pool due to random sampling, thereby reducing the number of operations and further optimizing the convergence performance.

Figure 5 depicts the fitting of the training curve tracking the reward function under different schemes. Specifically, Eq. (8) captures the calculation method of system utility, which is used to measure the performance of the two resource allocation schemes used in this work. As shown in the figure, the red curve gets the highest reward, which means that our proposed combination of blockchain sharing and priority experience replay will have the best utility while converging quickly. Moreover, with the increase of events, the utility of the red curve has been significantly improved. This is because the system converges quickly and improves the utility of computing resources. As the number of elements in the state space increases, the environment becomes more complex, resulting in a decrease in convergence performance, so the ascent of the curve slows down as the plot increases. However, in the case of any number of features, the joint consideration of network resource and computing resource allocation can allocate more powerful edge computing servers for users, so the resource utilization rate has been significantly improved compared with traditional DRL resource allocation schemes.

6   Conclusions and Future Work

In this article, we proposed a novel joint resource allocation scheme based on sharing mechanism with blockchain for future Internet of Vehicles. We first executed the local deep reinforcement learning algorithm with a prioritized experience replay mechanism on each edge node. Then, each edge node shared its learning result with others via blockchain in order to optimize the joint resource allocation problem. Most importantly, we considered channel resources and computing resources as components of a joint resources allocation strategy. Considering the importance of the samples in the experience replay pool, we used a prioritized experience replay to accelerate the local training speed. This approach can significantly improve the training efficiency through the simulation results. Incorporation of intrusion detection systems will be considered in the future.

Acknowledgements. We gratefully acknowledge the financial support from the Natural Sciences and Engineering Research Council of Canada (NSERC) under Grants No. RGPIN-2020-06482.

References

1. Liu, J., Wan, J., Zeng, B., Wang, Q., Song, H., Qiu, M.: A scalable and quick-response software defined vehicular network assisted by mobile edge computing. IEEE Commun. Mag. **55**(7), 94–100 (2017)
2. Zhang, Z., Wang, R., Yu, F.R., Fu, F., Yan, Q.: QoS aware transcoding for live streaming in edge-clouds aided HetNets: an enhanced actor-critic approach. IEEE Trans. Veh. Technol. **68**(11), 11295–11308 (2019)
3. He, Y., Yu, F.R., Zhao, N., Yin, H., Boukerche, A.: Deep reinforcement learning (DRL)-based resource management in software-defined and virtualized vehicular ad hoc networks. In: Proceedings of the 6th ACM Symposium on Development and Analysis of Intelligent Vehicular Networks and Applications (DIVANet 2017), New York, NY, USA, 47C54 (2017)
4. Wang, P., Di, B., Zhang, H., Bian, K., Song, L.: Cellular V2X communications in unlicensed spectrum: harmonious coexistence with VANET in 5G systems. IEEE Trans. Wireless Commun. **17**(8), 5212–5224 (2018)
5. Luo, G., et al.: Software-defined cooperative data sharing in edge computing assisted 5G-VANET. IEEE Trans. Mob. Comput. **20**(3), 1212–1229 (2021)
6. Zhang, D., Yu, F.R., Yang, R.: Blockchain-based distributed software-defined vehicular networks: a dueling deep Q -learning approach. IEEE Trans. Cognit. Commun. Networking **5**(4), 1086–1100 (2019)
7. Aublin, P.-L., Mokhtar S.B., Quéma, V.: RBFT: redundant byzantine fault tolerance. In: 2013 IEEE 33rd International Conference on Distributed Computing Systems, pp. 297–306 (2013)
8. He, Y., Zhao, N., Yin, H.: Integrated networking, caching, and computing for connected vehicles: a deep reinforcement learning approach. IEEE Trans. Veh. Technol. **67**(1), 44–55 (2018)
9. Zhang, H., Xiao, Y., Bu, S., Niyato, D., Yu, F.R., Han, Z.: Computing resource allocation in three-tier IoT fog networks: a joint optimization approach combining stackelberg game and matching. IEEE Internet Things J. **4**(5), 1204–1215 (2017)
10. Tian, J., Han, Q., Lin, S.: Improved delay performance in VANET by the priority assignment. In: IOP Conference Series: Earth and Environmental Science, vol. 234, no. 1 (2019)
11. Schaul, T., Quan, J., Antonoglou, I., et al.: Prioritized experience replay. arXiv preprint arXiv:1511.05952 (2015)
12. Abadi, M., Agarwal, A., Barham, P., et al.: Tensorflow: large-scale machine learning on heterogeneous distributed systems. arXiv preprint arXiv:1603.04467 (2016)
13. Qiu, C., Yao, H., Wang, X., Zhang, N., Yu, F.R., Niyato, D.: AI-chain: blockchain energized edge intelligence for beyond 5G networks. IEEE Network **34**(6), 62–69 (2020)

Ranging of Confocal Endoscopy Probe Using Recognition and Optical Flow Algorithm

Haoxiang Yu[1], Yuhua Lu[2], and Qian Liu[1,2(✉)]

[1] Key Laboratory of Biomedical Engineering of Hainan Province, School of Biomedical Engineering, Hainan University, Haikou 570228, China
qliu@hainanu.edu.cn
[2] Wuhan National Laboratory for Optoelectronics, Huazhong University of Science and Technology, Wuhan, China

Abstract. Early diagnosis is of great significance for the treatment of GI diseases. Endoscopic tissue biopsy of the GI tract is the standard means to diagnose whether the tumor will become cancerous or to confirm the stage of lesions. Confocal endoscopy, which can provide cell-level resolution and realize non-invasive real-time optical biopsy, has attracted much attention in the field of clinical disease diagnosis. However, because of the small field of vision, it is difficult to locate the probe again, which affects the efficiency and learning cost of confocal endoscopy. Based on the honeycomb shape of the original image of the confocal endoscope, the scale of the image pixel corresponding to the real world is obtained through the parameters of the endoscopy probe and the fiber bundle that make up the probe. Yolov3 is used as a crypt recognition algorithm, which assists the optical flow algorithm to predict the pixel displacement of crypt. Finally, the distance and angle of the endoscope lens motion are obtained through the scale. The moving angle and distance of the endoscope can help locate the probe position of the endoscope lens and record the probe path of the endoscope lens. After the exploration is completed, the original exploration path can also be restored by distance and angle information, which helps to accurately locate the lesion again. The Yolov3 Recognition Network was trained with 200 confocal endoscope images of rat colon. The $map_{0.5}$ was 99.84%. Compared with many other optical flow methods, the DISflow optical flow algorithm is finally selected. After testing, the angle error of the algorithm is less than 4°, the distance error is less than 8%. This work can restore the exploration path of confocal endoscopy and improve the diagnostic efficiency of confocal endoscopy.

Keywords: Optical flow · Obeject detection · Confocal endoscopy

1 Introduction

Since the invention of medical devices, gastroenterologists have aimed to detect gastrointestinal diseases by microlesions or non-invasive optical imaging in vivo. Biopsy is the gold standard for detecting gastrointestinal disease [1]. However, the biopsy process is time consuming, and the gastrointestinal tract cannot be observed during biopsy.

Therefore, confocal endoscopes, which combine optical fiber bundles with confocal microscopy, can be employed to addresses this issue.

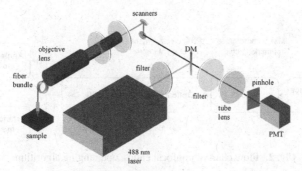

Fig. 1. Schematic diagram of confocal endoscope

The moving angle and distance of the confocal endoscope can assist to locate the exploration position of the endoscope probe, and record the exploration path of the endoscope probe. After the probe is complete, the original probe path is restored by distance and angle information. Thus, it allows to accurately locate the lesion again.

According to the nature of confocal endoscope, it is difficult to obtain the ground-truth optical flow from confocal endoscope images. Therefore, we choose the traditional optical flow method DISFlow [22]. Yolov3 [15] is employed as the crypt recognition algorithm in endoscope images. It uses crypt as markers in adjacent frames to measure the distance and angle of probe movement.

Current object detection algorithms usually comprise a backbone network, which is pre-trained on ImageNet as well as a Subnetwork and used to predict object categories and bounding boxes. The backbone network can be composed of VGG [18], ResNet [19], or DenseNet [20]. The Subnetworks for predicting object category and bounding box are divided into two types. First, a one-stage object detector—the most representative ones are YOLO [13–15], SSD [16], and RetinaNet [17]. Second, a two-stage object detector—the most representative algorithm is R-CNN [21].

The optical flow algorithm is applied to two images to describe the correspondence between the pixels of the two images. The classical optical flow method considers optical flow estimation as an optimization problem [2, 3]. For sequential image pairs, the optical flow field can maximize the smoothness and match the similarity of pixels. Moreover, there are references to recent deep learning methods to learn and estimate optical flow from examples of ground-truth image pairs through supervised training of deep neural networks [4–7]. However, it is difficult to obtain the flow field of the real world image. Thus, supervised learning generally uses artificially synthesized data [8, 9]. Although the optical flow method with supervised learning has achieved good results on synthetic datasets, the generalization of the supervised learning method is challenging when the optical flow scene to be predicted is significantly different from the training scene. There are also unsupervised optical flow methods [10–12] however, they require large data for training.

2   Principles and Methods

2.1   Algorithm Process

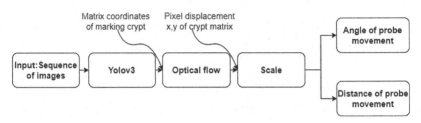

Fig. 2. Flow chart of confocal endoscope ranging algorithm

The input of algorithm is two images of adjacent frames. Yolov3 is used to obtain matrix coordinate information for all marked crypts in the first image. The DISFlow optical flow algorithm is used to predict the displacement of selected crypt in two consecutive frames. We obtained a 3-dimensional matrix (H $*$ W $*$ 2) from DISFlow. This 3D matrix records the lateral offset I_x and longitudinal offset I_y of the corresponding pixels from two frames. The moving angle and distance of the confocal endoscope probe in pixels/second can be obtained by calculating the I_x and I_y of all pixels in the labeled crypt matrix region. Finally, according to the unique honeycomb image of confocal endoscope, the images' real world scale and real world moving distance in millimeter can be obtained.

2.2   Dataset

In this study, we used a large field of view hose confocal endoscopic imaging system [23]. The rat colon tissue images were used in the experiment. In Yolov3, 200 colon tissue images of rats were selected as the dataset for training. We used labelimg to label crypt in rat colon tissue images. In the process of labeling, only the complete crypts in the image were labeled, and the incomplete crypts around the image were abandoned. In the proposed algorithm, we selected four different colonic data of rats. Four groups of colonic data were selected from each rat (each group contained two consecutive frames of images) for the experiment, and the final results were averaged.

2.3   Crypt Recognition Algorithm

In this study, Yolov3 algorithm was used as the crypt recognition algorithm. Yolov3 has the backbone network Darknet similar to ResNet, and can make multi-scale prediction. The coordinates of crypts can be obtained by using trained Yolov3 network. It can assist the optical flow to determine the direction. Furthermore, it can reduce the error of ranging distance and direction caused by the lens distortions of endoscope.

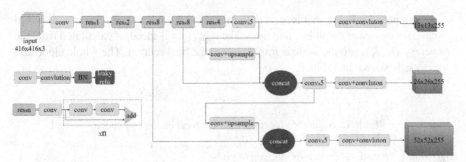

Fig. 3. Yolov3's backbone

2.4 Optical Flow

In this study, optical flow was used to obtain a two-dimensional velocity vector field. It represented the moving speed of the probe (Table 1). The optical flow can calculate the movement or displacement of image pixels between two consecutive frames of the image sequence. The classical optical flow formula is based on two assumptions [3]:

1. Brightness constant. Assume that the pixel intensity of the object between two consecutive frames does not change:

$$Ix + dx, y + dy, t + dt = Ix, y, t \tag{1}$$

   where I(x, y, t) represents the brightness (gray value) of the pixel of coordinate (x ,y) at time t.
2. Spatial smoothness consistency. By employing Eq. (1) alone to estimate the two unknown components of the velocity, u and v, does not yield a unique solution. This is called the aperture problem for optical flow. Usually, an additional smoothing term $E_{regu}(u, v)$ is used to solve this problem. It is assumed that adjacent pixels tend to have similar motions:

$$E_{regu}(u, v) = \iint_{\Omega} \left(|\nabla u|^2 + |\nabla v|^2 \right) dxdy. \tag{2}$$

   Here, ∇ represents the gradient operator.
   After obtaining the optical flow matrix predicted by the optical flow method, the moving distance and moving angle of the probe can be obtained through the following two equations:

$$angle = \arctan \frac{I_y}{I_x} \tag{3}$$

   The angle can be obtained by a trigonometric formula. I_x is the average pixel displacement in the horizontal direction. I_y is the average pixel displacement in the vertical direction.

$$d_{real} = v_{pixel} * d_{pixel} * fps * t \tag{4}$$

The probe moving distance is determined by the pixel speed, pixel-real world scale, frame rate of the video, and time. The v_{pixel} (pixel speed) is calculated from the average (I_x, I_y), and the scale is introduced in the next section. The whole algorithm process is shown in Table 1:

Table 1. Algorithm for ranging of confocal endoscopy probe

Algorithm 1. Proposed approach for ranging of probe
Input: Sequence of images I_t, $t = 1, 2, \ldots, n$, The selected crypt coordinates
Output: Moving distance and angle of probe 1. For each consecutive image pair (I_t, I_{t+1}), and the coordinates of the Selected crypt
2. The optical flow field ((I_x, I_y), -1)of each pixel in the matrix is calculated by optical flow
3. Calculate the average (I_x, I_y)offset of the matrix
4. The moving angle of the probe can be obtained by Eq. (3)
5. The moving distance of the probe can be obtained by Eq. (4)

2.5 Scale Calculation Method

In confocal endoscope imaging system, the probe composed of optical fiber bundle is an important medium for excitation light transmission and fluorescence signal collection. It ensures that the system has sufficient high resolution and large imaging field of view in the imaging area. However, the optical fiber bundle is composed of multiple fibers, and its imaging structure results in limited resolution [23, 24]. It images are displayed in a honeycomb pattern. Based on this imaging method, we can count out the pixels in one fiber circle, and then obtain a real-world scale of the image.

By using the radius parameter r of a fiber circle in the endoscope, the area of the effective image circle can be obtained as πr^2, assuming $\pi \approx 3.14$. The area of an effective image circle is expressed as:

$$S_{image\ circle} = \pi r^2 \tag{5}$$

The area of a fiber circle is approximately 16 pixels on the image (Fig. 4(B)), thus the area of a pixel is as follows:

$$S_{pixel} = \frac{S_{image\ circle}}{16} \tag{6}$$

A pixel is approximated as a small circle. The diameter of a pixel is:

$$d_{pixel} = 2 \times \sqrt{S_{pixel}/\pi} \tag{7}$$

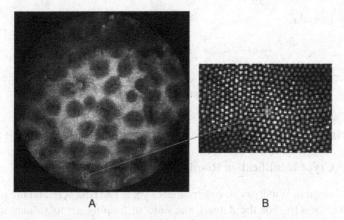

Fig. 4. (A) Honeycomb confocal endoscope image; (B) enlarged view of area image

2.6 Experimental Reference Object

Crypts are structures of the gut. It is home to a mass proliferation of epithelial cells, which can be easily observed with confocal endoscopes [25]. Owing to the randomness of the operator in the actual use of confocal endoscope, and because there is no function of recording displacement currently, it is difficult to obtain the real displacement of the probe. In this study, we calculated the offset of the crypt centroid in two consecutive frames to obtain the moving distance and angle of the probe. As a reference standard, it was used to judge the error of the angle and distance predicted by the optical flow method.

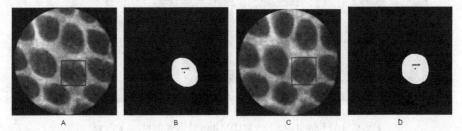

Fig. 5. (A) Frame 1; (B) Segmentation crypt of Frame 1; (C) Frame 2; (D) Segmentation crypt of Frame 2

We obtained selected crypts by binarization and segmentation. The crypt is approximated as a polygon. Two centroid points $p_{centroid1}(x_{centroid1}, y_{centroid1})$, $p_{centroid2}(x_{centroid2}, y_{centroid2})$ can be obtained by centroid formula.

Angle formula:

$$angle_{true} = \arctan \frac{y_{centroid2} - y_{centroid1}}{x_{centroid2} - x_{centroid1}} \tag{8}$$

Distance formula:

$$d_{true} = d_{pixel} * d_{centroid} * fps * Time \tag{9}$$

The reference-moving angle $angle_{true}$, and the reference moving distance d_{true} of probe can be obtained by the angle formula 8 and distance formula 9.

3 Experimental Results

3.1 Yolov3 Crypt Identification Results

Training was done in Windows 10 environment by a GTX1650(4G), and the image was resized to 416×416. For the dataset, the ratio of training set to validation set was 9:1. Pytorch and cuda versions were 1.10.0 and 11.3. The learning rate was $1e^{-4}$. The optimizer was Adam and epoch was 500. In Yolov3 algorithm, the map can reach 99.84% because the network structure is relatively simple. Thus, the graphics card computing power requirements are also low. The crypt coordinates in the image were obtained by Yolov3. The moving angle and moving distance were obtained by crypt coordinates and optical flow (Figs. 1, 2, 3, 5, and 6).

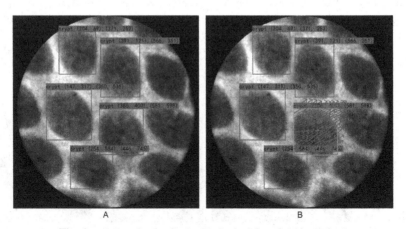

Fig. 6. (A) Result of yolov3; (B) Optical flow field by Yolov3

3.2 Endoscope Ranging Result by Optical Flow and Yolov3

In this study, we selected five traditional optical flow methods: DeepFlow [26], DISFlow [22], Farneback [27], PCA_Flow [28] and Sparse_to_dense [29]. We combined the methods with Yolov3 to estimate the moving distance and angle of the probe. The scale was composed of the fiber bundle parameters of the probe: outer diameter of fiber bundle was 1200 μm, image circle diameter was 1125 μm, and a Fiber diameter was 4.38 μm. The scale of real world-image pixel was 1.095 μm. Four groups of different endoscope data were tested by the five traditional optical flow methods. Another 4 groups of data were selected from each group of endoscopy data. The error of each optical flow method was obtained by averaging the four groups of data, and the traditional optical flow method with the best performance was obtained by summarizing at the end.

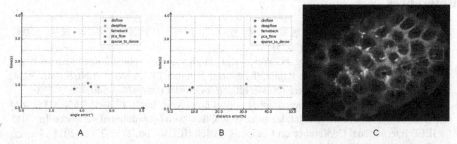

Fig. 7. (A) Mean angle error of five optical flow methods; (B) Mean distance error of five optical flow methods; (C) Probe trajectory

Figure 7 illustrates that DeepFlow, DISFlow, and Sparse_to_dense are the three algorithms with better performance in predicting angle and distance. Their angle and distance errors can be controlled at approximately 4° and 8%, respectively. In prediction time, DISFlow and Sparse_to_dense are the two algorithms with better performance, the prediction time is within one second. Because the angle error and running time performance of DISFlow are better than those of the Sparse_to_dense. In this study, we chose DISFlow optical flow algorithm. Figure 7(C) is a Mosaic of 18 pictures. The displacement of the selected crypt in every two consecutive frames can be obtained by the optical flow method combined with the crypt recognition algorithm. Thereafter, the path of the selected crypt in the lens can be drawn, which can also be the path of the probe in the stomach.

4   Conclusion

This study presents a method that employs Yolov3 as a crypt recognition algorithm and an optical flow algorithm to estimate the probe movement angle and distance of the confocal endoscope. The results showed that the total error of Yolov3 crypt recognition algorithm combined with DISFlow optical flow algorithm in estimating the lens moving distance was only 8%. The error on different data was approximately 10%, not exceeding 15%. The Angle error was only approximately 4°. The estimated time was less than one second. Although the results are good, more research is necessary to evaluate the performance of the algorithm in the context of unclear camera images. However, these preliminary results show that this method can accurately measure the angle and distance of the moving probe of the endoscope within an error range. It can assist doctors to use confocal endoscope to locate the probe exploration position, and restore the exploration path through the obtained information after the exploration. When doctors use confocal endoscopes, more information can be obtained, and more accurate localization and diagnosis of lesions can be established.

Acknowledgments. Funding. Hainan Province Key Science and Technology Project (ZDKJ202006). (Supported by Major Special Science and Technology Project of Hainan Province, ZDKJ202006).

References

1. 王家福, 杨敏, 杨莉, 张云, 袁菁, 刘谦, 侯晓华, 付玲. 用于细胞成像的共聚焦内窥镜. Engineering **1**(03), 152–171 (2015)
2. Brox, T., Bruhn, A., Papenberg, N., Weickert, J.: High accuracy optical flow estimation based on a theory for warping. In: Pajdla, T., Matas, J. (eds.) ECCV 2004. LNCS, vol. 3024, pp. 25–36. Springer, Heidelberg (2004). https://doi.org/10.1007/978-3-540-24673-2_3
3. Horn, B.K., Schunck, B.G.: Determining optical flow. Artif. Intell. **17**(1–3), 185–203 (1981)
4. Dosovitskiy, A., et al.: FlowNet: learning optical flow with convolutional networks. In: 2015 IEEE International Conference on Computer Vision (ICCV), pp. 2758-2766 (2015). https://doi.org/10.1109/ICCV.2015.316
5. Ilg, E., Mayer, N., Saikia, T., Keuper, M., Dosovitskiy, A., Brox, T.: Flownet 2.0: evolution of optical flow estimation with deep networks. In: CVPR (2017)
6. Sun, D., Yang, X., Liu, M.Y., Kautz, J.: PWC-Net: CNNs for optical flow using pyramid, warping, and cost volume. In: CVPR (2018)
7. Teed, Z., Deng, J.: RAFT: recurrent all-pairs field transforms for optical flow. In: Vedaldi, A., Bischof, H., Brox, T., Frahm, J.-M. (eds.) ECCV 2020. LNCS, vol. 12347, pp. 402–419. Springer, Cham (2020). https://doi.org/10.1007/978-3-030-58536-5_24
8. Mayer, N., et al.: A large dataset to train convolutional networks for disparity, optical flow, and scene flow estimation. In: CVPR (2016)
9. Butler, D.J., Wulff, J., Stanley, G.B., Black, M.J.: A naturalistic open source movie for optical flow evaluation. In: Fitzgibbon, A., Lazebnik, S., Perona, P., Sato, Y., Schmid, C. (eds.) ECCV 2012. LNCS, vol. 7577, pp. 611–625. Springer, Heidelberg (2012). https://doi.org/10.1007/978-3-642-33783-3_44
10. Jonschkowski, R., Stone, A., Barron, J.T., Gordon, A., Konolige, K., Angelova, A.: What matters in unsupervised optical flow. In: Vedaldi, A., Bischof, H., Brox, T., Frahm, J.M. (eds.) ECCV 2020. LNCS, vol. 12347, pp. 557–572. Springer, Cham (2020). https://doi.org/10.1007/978-3-030-58536-5_33
11. Meister, S., Hur, J., Roth, S.: UnFlow: unsupervised learning of optical flow with a bidirectional census loss. In: Proceedings of the AAAI Conference on Artificial Intelligence, vol. 32, no. 1 (2018). https://doi.org/10.1609/aaai.v32i1.12276
12. Luo, K., Wang, C., Liu, S., Fan, H., Wang, J., Sun, J.: UPFlow: upsampling pyramid for unsupervised optical flow learning In: IEEE/CVF Conference on Computer Vision and Pattern Recognition (CVPR), pp. 1045 1054 (2021). https://doi.org/10.1109/CVPR46437.2021.00110
13. Redmon, J., Divvala, S., Girshick, R., Farhadi, A.: You only look once: unified, real-time object detection. In: Proceedings of the IEEE Conference on Computer Vision and Pattern Recognition (CVPR), pp. 779–788 (2016)
14. Redmon, J., Farhadi, A.: YOLO9000: better, faster, stronger. In: Proceedings of the IEEE Conference on Computer Vision and Pattern Recognition (CVPR), pp. 7263–7271 (2017)
15. Redmon, J., Farhadi, A.: YOLOv3: An incremental improvement. arXiv preprint arXiv:1804.02767 (2018)
16. Liu, W., et al.: SSD: single shot multibox detector. In: Leibe, B., Matas, J., Sebe, N., Welling, M. (eds.) ECCV 2016. LNCS, vol. 9905, pp. 21–37. Springer, Cham (2016). https://doi.org/10.1007/978-3-319-46448-0_2

17. Lin, T.Y., Goyal, P., Girshick, R., He, K., Dollár, P.: Focal loss for dense object detection. In: Proceedings of the IEEE International Conference on Computer Vision (ICCV), pp. 2980–2988 (2017)

18. Simonyan, K., Zisserman, A.: Very deep convolutional networks for large-scale image recognition. arXiv preprint arXiv:1409.1556 (2014)

19. He, K., Zhang, X., Ren, S., Sun, J.: Deep residual learning for image recognition. In: Proceedings of the IEEE Conference on Computer Vision and Pattern Recognition (CVPR), pp. 770–778 (2016)

20. Huang, G., Liu, Z., Van Der Maaten, L., Weinberger, K.Q.: Densely connected convolutional networks. In: Proceedings of the IEEE Conference on Computer Vision and Pattern Recognition (CVPR), pp. 4700–4708 (2017)

21. Girshick, R., Donahue, J., Darrell, T., Malik, J.: Rich feature hierarchies for accurate object detection and semantic segmentation. In: Proceedings of the IEEE Conference on Computer Vision and Pattern Recognition (CVPR), pp. 580–587 (2014)

22. Kroeger, T., Timofte, R., Dai, D., Van Gool, L.: Fast optical flow using dense inverse search. In: Leibe, B., Matas, J., Sebe, N., Welling, M. (eds.) ECCV 2016. LNCS, vol. 9908, pp. 471–488. Springer, Cham (2016). https://doi.org/10.1007/978-3-319-46493-0_29

23. 李华. 大视场共聚焦内窥成像方法研究. 华中科技大学 (2021)

24. 王家福. 近红外共聚焦内窥成像方法研究. 华中科技大学 (2018)

25. Clevers, H.: The intestinal crypt, a prototype stem cell compartment. Cell **154**(2), 274–284 (2013). https://doi.org/10.1016/j.cell.2013.07.004. PMID: 23870119

26. Weinzaepfel, P., Revaud, J., Harchaoui, Z., Schmid, C.: DeepFlow: large displacement optical flow with deep matching. In: 2013 IEEE International Conference on Computer Vision, pp. 1385–1392 (2013). https://doi.org/10.1109/ICCV.2013.175

27. Farnebäck, G.: Two-Frame motion estimation based on polynomial expansion. In: Bigun, J., Gustavsson, T., (eds) Image Analysis. SCIA 2003. Lecture Notes in Computer Science, vol 2749 (2003). Springer, Berlin, Heidelberg. https://doi.org/10.1007/3-540-45103-X_50

28. Wulff, J., Black, M.J.: Efficient sparse-to-dense optical flow estimation using a learned basis and layers. In: 2015 IEEE Conference on Computer Vision and Pattern Recognition (CVPR), pp. 120-130 (2015). https://doi.org/10.1109/CVPR.2015.7298607

29. OpenCV: Optical Flow Algorithms

Deployment Strategy of Highway RSUs for Vehicular Ad Hoc Networks Considering Accident Notification

Yufeng Chen[1], Yingkui Ma[1], Zhengtao Xiang[1]([⊠]), Lei Cai[2], Yu Zhang[1], and Hanwen Cao[1]

[1] School of Electrical and Information Engineering, Hubei University of Automotive Technology, Shiyan 442002, China
xztcyf@163.com
[2] Beijing GOTEC ITS Technology Co. Ltd., Beijing 100088, China

Abstract. In the highway scenario, the deployment of RSUs to improve the delivery delay of accident notification information is studied. In this paper, a theoretical analysis model is established by analyzing the relationship between the delivery delay of accident notification information and the deployment of RSUs in the Vehicular Ad Hoc Networks (VANETs). The model assumes that two adjacent RSUs are connected, and considers the impact of accident probability, vehicle node density, and speed on the delivery delay of accident notification information under different deployment distances. In addition, the model also considers secondary inter-cluster communication with vehicles in different driving directions from the accident vehicle, so as the accident notification information can be transmitted to the vehicle cluster with same direction of accident vehicle and closer to the RSUs. To validate the performance of our proposed model, this paper uses MATLAB to solve the model numerically. Compared with the existing model that assumes that two adjacent RSUs are not connected and the secondary inter-cluster communication are not considered, the delivery delay of the accident notification information of our proposed model can be reduced by 66% from the existing model.

Keywords: RSU deployment · VANETs · Accident notification information delivery delay · Secondary inter-cluster communication

1 Introduction

The rapid increase in road traffic accidents (RTAs) has caused massive property losses and casualties in different countries [1]. According to a study by the World Economic Forum in April 2016, the number of cars expected to double globally by 2040 will place more pressure on road transport infrastructure [2]. In addition, the World Health Organization (WHO) predicts that RTAs will become the seventh leading cause of death by 2030 [3]. In RTAs, the probability of an accident occurring within 100 km of a highway is about 4 times that of an ordinary road [4]. Since the highway is a linear fully enclosed road, if the accident notification is not made to the rear vehicle in time after the initial accident, the rear vehicle will not brake and collided with the accident vehicle in time, resulting in a secondary accident. To reduce the probability of secondary road traffic safety accidents and ensure road traffic safety, Vehicle Ad hoc Networks (VANETs) have been developed around the world. Because the vehicle itself moves at a high speed, the topology of VANETs is highly dynamic, which leads to intermittent connectivity of the network, thus increasing the delay of accident information collection and transmission. To solve this problem, a roadside unit (RSU) is introduced into the VANETs. The deployment of RSU can improve the connectivity of the network, thereby greatly improving the real-time performance of accident information collection and transmission [5]. However, the single-unit cost and deployment cost of RSUs is extremely high [6], and it is impossible to deploy on a large scale to make the entire network reach a fully connected state. It is necessary to reasonably determine the deployment distance of RSUs.

We study the delay analysis of accident notification information transmission considering RSU deployment in intermittently connected VANETs. Considering the interconnection between two adjacent RSUs, we establish a theoretical analysis mathematical model to study the relationship between the average delay of the delivery of accident notification information transmission and the distance between RSU deployments.

The work done in this paper is as follows

(1) We studied the deployment of RSUs to improve the delivery delay of accident notification information in a highway scenario. We consider a direct connection between two adjacent RSUs and establish a theoretical analysis model by analyzing the relationship between the delivery delay of accident notification information in VANETs and the deployment distance of interconnected RSUs.
(2) We propose a secondary communication method between vehicle clusters. This method considers secondary inter-cluster communication with vehicles in different driving directions from the accident vehicle, so as the accident notification information can be transmitted to the vehicle cluster with same direction of accident vehicle and closer to the RSUs.
(3) We numerically solve the model using MATLAB and compare it with an existing model that assumes that two adjacent RSUs are not interconnected to determine the validity of our proposed model.

The organizational structure of this paper is as follows: Sect. 2 will introduce related work. Section 3 describes our network scenario. Section 4 will give the model of accident notification information transmission delay. In Sect. 5, we will simulate the theoretical model and analyze the numerical results. Section 6 is our conclusion.

2   Related Work

References [5, 7–14] studied the highway RSU deployment. Liang et al. [7] pointed out that measurement errors and inference errors should be considered in RSU deployment, defined RSU deployment as a bi-objective nonlinear binary integer programming problem, and proposed an ε-constraint solving method. Considering the proportion and time threshold of vehicles, Silva et al. [8] proposed a Gamma deployment strategy to define the location and number of RSUs required to provide specific coverage. Ge et al. [9] used an improved K-means clustering algorithm to cluster vehicles and obtained the deployment distance of RSU by analyzing the relationship between RSU transmission range, vehicle density, vehicle connectivity rate, and average vehicle cluster length.

The above references only studied the setting of RSU deployment distance under normal traffic flow but did not consider the setting of RSU deployment distance in case of traffic accidents on the road. Sou et al. [10] analyzed and quantified the improvement of VANETs connectivity when deploying a limited number of RSUs, and studied the routing pros and cons of incident advertisement messaging in this enhanced VANETs environment. Wisitpongphan et al. [11] proposed a "store-carry-forward" strategy to study the accident notification message delivery delay using a statistical model extracted from measured data on the I-80 freeway in California. However, this strategy only considered the vehicle-to-vehicle message delivery delay but did not consider the vehicle-to-RSU delivery delay. Wang et al. [5, 12–14] further improved the model proposed by Wisitpongphan et al. [11], and proposed a new theoretical analysis model of the relationship between accident notification message delivery delay and RSU deployment distance, which assumes that two adjacent RSUs are not directly connected, and integrates the vehicle speed, vehicle density, accident location and two adjacent RSUs with different. The deployment distance between two adjacent RSUs is given by considering the relationship between vehicle speed, vehicle density, accident location, and different deployment distances of two adjacent RSUs. The experimental results show that the theoretical analysis model can be used to determine the maximum allowed deployment interval when two adjacent RSUs are not connected to each other. However, there is still room for further enhancement of the model as follows: 1) The model does not consider the direct connection between two adjacent RSUs; 2) In this model, only one inter-cluster communication is carried out, and the accident notification information is transmitted to the vehicle clusters in different driving directions of the accident vehicle. The secondary inter-cluster communication with vehicles in different driving directions is not considered, and the accident notification information is passed to the vehicle cluster in the same driving direction as the accident vehicle and closer to the RSU.

3 Network Scenarios

We stipulate that all vehicles are configured with the onboard unit (OBU), and global navigation satellite system (GNSS), and periodically broadcast beacon messages to surrounding vehicles. After receiving a neighbor beacon message, each vehicle must compare its GNSS information to determine whether it is traveling in the same direction as it is traveling. By constantly exchanging beacons, each vehicle maintains its list of lead, follow, and opposite one-hop neighbors. If vehicle nodes in the same driving direction can communicate in one or more hops, these vehicle nodes can be clustered into a cluster. A cluster that contains only one vehicle is called a single cluster, as shown in Fig. 1.

We consider a highway scene with sparse vehicle nodes, two RSUs are deployed at both ends of the road as shown in Fig. 1. Assume that the RSU located upstream of the eastbound vehicle is RSU1, which is at position 0. Assume that the RSU located downstream of the east-bound vehicle is RSU2, which is at position d. The RSU transmission range is R_u, and the vehicle node transmission range is R_v. We assume that $R_u = R_v$. RSU1 and RSU2 are connected by cables, but the transmission range does not overlap each other, that is, $d \gg 2R_u$. Vehicle nodes traveling east and west randomly arrive at both ends of the road with the arrival rates of λ_e and λ_w, and the arrival distance obeys an exponential distribution [5]. Assuming that the speed of the vehicle is stochastic and obeys a truncated normal distribution [15], the average speeds of vehicle nodes traveling east and west are defined as v_e and v_w. If the accident vehicle is located in the transmission range of the two RSUs (i.e., $x \in [0, R_u]$ or $x \in [d - R_u, d]$), the accident vehicle can directly notify the accident information RSU. In this case, the time spent is far less than the "store-carry-forward" between vehicle nodes. The delay caused can be ignored. Therefore, we only consider the case where the accident information occurs in the blind area of the deployment of two RSUs (i.e., $x \in [R_u, d - R_u]$). When the accident occurs, the accident vehicle will select the appropriate forwarding node to forward the accident information until the event information is transmitted to RSU1 or RSU2. The average information delivery time required from the time of accident occurrence to the time when RSU1 or RSU2 receives accident information is defined as the average accident notification information delivery delay. In the highway scenario with sparse vehicle nodes, the probability of vehicle nodes competing for the channel at the MAC layer is very small, and the delay generated by channel competition in the MAC layer is also very low compared to the delay generated by "store-carry-forward", so it can be ignored. In addition, we ignore the delay in the transmission of accident notification information by cable between two adjacent RSUs.

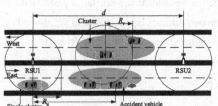

Fig. 1. Network scenario

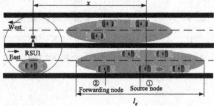

Fig. 2. Node 2 becomes a forwarding node (Case 1)

4 Modeling of Information Delivery Delay

We assume that the accident occurred in a vehicle traveling eastward, the probability density function (PDF) of the distance x from the accident location to the RSU1 location is denoted as $f(x)$, and $f(x)$ obeys a uniform distribution. Due to the randomness of the accident location, the time required by the forwarding node to transmit the accident information to RSU1 or RSU2 is not equal. Since we assume that two adjacent RSUs are connected by cables, the average delivery delay of the accident notification information is taken as the short time required for the forwarding node to forward the accident notification information to RSU1 or RSU2. Based on the above considerations, the average accident notification information delivery delay can be approximated by

$$T = \int_{R_u}^{d-R_u} \min\{T_{RSU1}, T_{RSU2}\} \cdot f(x) dx, \tag{1}$$

where T is the average accident notification information delivery delay, and T_{RSU1} and T_{RSU2} are the time required for the forwarding node to transmit the accident notification information to RSU1 and RSU2. T_{RSU1} and T_{RSU2} are analyzed separately below. The main symbol definitions used in this paper are shown in Table 1.

Table 1. Main symbol definitions used in this paper.

Parameters	Values
l_e	The length of a cluster of eastbound vehicles
l_w	The length of a cluster of westbound vehicles
$E[l_e]$	The average l_e
$E[l_w]$	The average l_w
N_e	The number of vehicles in a cluster of eastbound vehicles
N_w	The number of vehicles in a cluster of westbound vehicles
$f_e(N_e)$	The probability mass function (PMF) of N_e
$f_w(N_w)$	The probability mass function (PMF) of N_w
$E[d_v^e]$	The average distance between the two vehicles is in a cluster of eastbound
$E[d_v^w]$	The average distance between the two vehicles is in a cluster of westbound
$E[d_c^e]$	The average distance between two adjacent eastbound vehicle clusters
$E[d_c^w]$	The average distance between two adjacent westbound vehicle clusters

4.1 Analysis of T_{RSU1}

T_{RSU1} analysis is shown in Fig. 3. According to [14], we can assume that the accident vehicle is located in the middle of the vehicle cluster. Because the deployment location of RSU1 is located upstream of the accident vehicle, the accident vehicle selects the vehicle node (i.e., node 2) on the westernmost side of the vehicle cluster where it is located to be the forwarding node of the accident notification information. When node 2 is located in the transmission range of RSU1, node 2 forwards the accident notification information to RSU1. At this time, the delay caused by the direct forwarding of information is much smaller than that of storage carrying forwarding, which can be ignored. Therefore, it is only necessary to study the situation of node 2 outside the RSU1 transmission range (i.e., Case 1), as shown in Fig. 2. According to [12], we define the probability of this happening is defined as p_1.

$$p_1 = 1 - \left(1 - e^{-\lambda_e \cdot R_v}\right)^{k_1}, \tag{2}$$

where $k_1 = 2 \cdot (x - R_u + R_v)/E[d_v^e]$. According to [14], we have

$$E[d_v^e] = \frac{1}{\lambda_e} - \frac{R_v \cdot e^{-\lambda_e \cdot R_v}}{1 - e^{-\lambda_e \cdot R_v}}.$$ (3)

According to [11], we define the conditional expectation E_1 when half cluster length $l_e/2$ satisfies the condition $l_e/2 - R_v < x - R_u$.

$$E_1 = 0.5 E[d_v^e] \cdot \frac{1 - (1 - p_1) \cdot \left(k_1 \cdot e^{-\lambda_e \cdot R_v} + 1\right)}{p_1 \cdot e^{-\lambda_e \cdot R_v}}.$$ (4)

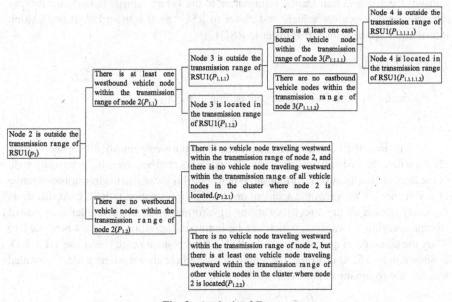

Fig. 3. Analysis of T_{RSU1}

(1) Case 1.1

According to [12], we define the probability of this happening as $p_{1.1}$.

$$p_{1.1} = 1 - e^{-2\lambda_w \cdot R_v}.$$ (5)

At this time, the westmost vehicle node (i.e., node 3) in the vehicle cluster where the westbound vehicle node is located becomes the forwarding node of the accident notification information. When node 3 is located in the transmission range of RSU1, node 3 forwards the accident notification information to RSU1. At this time, the delay caused by the direct forwarding of information is much smaller than that of storage carrying forwarding, which can be ignored. Therefore, it is only necessary to study the situation of node 3 outside the RSU1 transmission range (i.e., case 1.1.1), as shown in Fig. 4. According to [12], the probability of this happening can be calculated as

$$p_{1.1.1} = 1 - (1 - e^{-\lambda_w \cdot R_v})^{k_{1.1.1}}, \tag{6}$$

where $k_{1.1.1} = 2 \cdot (x - E_1 + R_v)/E[d_v^w]$. According to [14], $E[d_v^w]$ can be calculated as

$$E[d_v^w] = \frac{1}{\lambda_w} - \frac{R_v \cdot e^{-\lambda_w \cdot R_v}}{1 - e^{-\lambda_w \cdot R_v}}. \tag{7}$$

For this case, Wang et al. [5, 12–14] did not continue processing, but chose node 3 to continue carrying the accident notification information to RSU1. We propose a method for secondary communication between vehicle clusters. This method uses node 3 to transmit the accident notification information to the vehicle cluster in the same driving direction as the accident vehicle and closer to RSU1 so that the accident notification information can be transmitted faster to RSU1.

1) Case 1.1.1.1

We define the probability of this happening as $p_{1.1.1.1}$.

$$p_{1.1.1.1} = 1 - e^{-2\lambda_e \cdot R_v}. \tag{8}$$

At this time, the vehicle node (i.e., node 4) on the westernmost side of the vehicle cluster where the eastward traveling vehicle node is located becomes the forwarding node of the accident notification information. When node 4 is located in the transmission range of RSU1, node 4 forwards the accident notification information to RSU1. At this time, the delay caused by the direct forwarding of information is much smaller than that of storage carrying forwarding, which can be ignored. Therefore, it is only necessary to study the situation of node 4 outside the RSU1 transmission range (i.e., Case 1.1.1.1), as shown in Fig. 5. At this time, the length of the vehicle cluster where node 4 is located satisfies the following conditions:

$$l_e - 2R_v < x - E_1 - E[d_c^e] + R_v.$$

We define the probability of this happening as $p_{1.1.1.1.1}$.

$$p_{1.1.1.1.1} = \Pr\{l_e < x - E_1 - E[d_c^e] + 3R_v\}. \tag{9}$$

According to [14], $E[d_c^e]$ can be calculated by

$$E[d_c^e] = R_v + \lambda_e^{-1}, \tag{10}$$

and according to [12], the PMF of N_e:

$$f_e(N_e) = e^{-\lambda_e \cdot R_v} \cdot (1 - e^{-\lambda_e \cdot R_v})^{N_e-1}. \tag{11}$$

Therefore, the vehicle cluster length l_e can be approximated by the equation. Given the probability mass function obeyed by $E[d_v^e]$, $E[d_c^e]$ and N_e, $p_{1.1.1.1.1}$ can be calculated as

$$p_{1.1.1.1.1} = \Pr\{l_e < x - E_1 - E[d_c^e] + 3R_v\}$$

$$= \Pr\{N_e \cdot E[d_v^e] < x - E_1 - E[d_c^e] + 3R_v\}$$

$$= \Pr\{N_e < \frac{x - E_1 - E[d_c^e] + 3R_v}{E[d_v^e]}\}$$

$$= \Pr\{N_e < k_{1.1.1.1.1}\} = \sum_{n_e=1}^{k_{1.1.1.1.1}} f_e(N_e)$$

$$= 1 - (1 - e^{-\lambda_e \cdot R_v})^{k_{1.1.1.1.1}}, \tag{12}$$

where $k_{1.1.1.1.1} = (x - E_1 - E[d_c^e] + 3R_v)/E[d_v^e].$

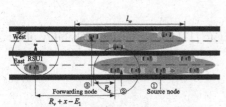

Fig. 4. Node 3 becomes a forwarding node (Case 1.1.1)   **Fig. 5.** Node 4 becomes a forwarding node (Case 1.1.1.1)

At this time, the deployment position of RSU1 is located upstream of node 4. Node 4 cannot carry the accident notification information to RSU1, and the accident notification information will continue to be carried by node 3 to RSU1. According to [12], the accident notification information delivery delay can be approximated by

$$T_{1.1.1.1.1} = (x - R_u - E_1 + 2R_v - E\left[\frac{l_w}{2} | \frac{l_w}{2} < x - E_1 + R_v\right]/2)/v_w, \tag{13}$$

where $E\left[\frac{l_w}{2} | \frac{l_w}{2} < x - E_1 + R_v\right] = 0.5E[d_v^w] \cdot \frac{1-(1-p_{1.1.1})\cdot(k_{1.1.1}\cdot e^{-\lambda_w \cdot R_v}+1)}{p_{1.1.1}\cdot e^{-\lambda_w \cdot R_v}}.$

2) Case 1.1.1.2

We define the probability of this happening as $p_{1.1.1.2}$.

$$p_{1.1.1.2} = 1 - p_{1.1.1.1}. \tag{14}$$

At this time, the accident notification information needs to be stored and carried by node 3 and passed to the RSU1. The calculation expression of the delay $T_{1.1.1.2}$ required for the transmission of the accident notification information is the same as that of $T_{1.1.1.1.1}$.

(2) Case 1.2

We define the probability of this happening as $p_{1.2}$.

$$p_{1.2} = 1 - p_{1.1}. \tag{15}$$

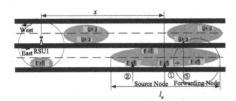
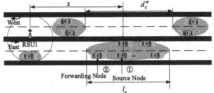

Fig. 6. Node 5 becomes a forwarding node (Case 1.2.1)

Fig. 7. Node 2 becomes a forwarding node (Case 1.2.2)

1) Case 1.2.1

Case 1.2.1 is shown in Fig. 6. According to [12], we define the probability of this happening as $p_{1.2.1}$.

$$p_{1.2.1} = (1 - e^{-\lambda_e \cdot R_v})^{k_{1.2.1}}, \tag{16}$$

where $k_{1.2.1} = (E[d_c^w] - 2R_v)/E[d_v^e]$.

At this time, node 2 directly transmits the accident notification information to the vehicle node (i.e., node 5) traveling westward, and the delay caused by the direct transmission of the accident notification information is negligible. Node 5 becomes a new forwarding node and transmits the accident notification information to RSU1. According to [5], the distance between node 5 and RSU1 is at most $x + 0.5E[l_e]$. Therefore, the accident notification information delivery delay is approximated by:

$$T_{1.2.1} = (x + E[l_e]/2)/v_w. \tag{17}$$

According to [14], we have

$$E[l_e] = (e^{\lambda_e \cdot R_v} - 1) \cdot \left(\frac{1}{\lambda} - \frac{R_v \cdot e^{-\lambda_e \cdot R_v}}{1 - e^{-\lambda_e \cdot R_v}} \right). \tag{18}$$

2) Case 1.2.2

Case 1.2.2 as shown in Fig. 7. We define the probability of this happening as $p_{1.2.2}$.

$$p_{1.2.2} = 1 - p_{1.2.1}. \tag{19}$$

At this time, the accident notification information will be stored and carried by node 2 until the vehicle node traveling westward is encountered, and the accident notification information will be forwarded to the vehicle node traveling westward. The vehicle node traveling westward becomes a new forwarding node and transmits the accident notification information to RSU1. Among the vehicle nodes traveling westward, the vehicle node closest to node 2 may become a new forwarding node, and the distance between this node and RSU1 is at least $x + 0.5E[d_c^w]$. According to [12], the accident notification information delivery delay is approximated by:

$$T_{1.2.2} = (x + E[d_c^w]/2)/v_w. \tag{20}$$

In summary, we have

$$T_{RSU1} = p_1 \cdot [p_{1.1} \cdot p_{1.1.1} \cdot (p_{1.1.1.1} \cdot p_{1.1.1.1.1} \cdot T_{1.1.1.1.1} + p_{1.1.1.2} \cdot T_{1.1.1.2})$$
$$+ p_{1.2} \cdot (p_{1.2.1} \cdot T_{1.2.1} + p_{1.2.2} \cdot T_{1.2.2})]. \tag{21}$$

4.2 Analysis of T_{RSU2}

T_{RSU2} analysis is shown in Fig. 8. Because the deployment location of RSU1 is located downstream of the accident vehicle, the accident vehicle can directly forward the accident notification information to the vehicle node (i.e., node 6) on the easternmost side of the vehicle cluster to become the forwarding node of the accident notification information. When node 6 is located in the transmission range of RSU2, the accident notification information will be forwarded from node 6 to RSU2. At this time, the delay caused by the direct forwarding of information is much smaller than that of storage carrying forwarding, which can be ignored. Therefore, it is only necessary to study the case where node 6 is outside the RSU2 transmission range (i.e., Case 2), as shown in Fig. 9. At this time, the length of the vehicle cluster where node 6 is located satisfies the following conditions:

$$l_e/2 < d - x - R_u + R_v$$

According to [12], we define the probability of this happening as p_2.

$$p_2 = 1 - (1 - e^{-\lambda_e \cdot R_v})^{k_2}, \tag{22}$$

where $k_2 = 2 \cdot (d - x - R_u + R_v)/E[d_v^e]$.

We define the conditional expectation E_2 when half cluster length $l_e/2$ satisfies the condition $l_e/2 < d - x - R_u + R_v$.

$$E_2 = E[d_v^e] \cdot E[N_e | N_e < k_2]/2, \tag{23}$$

where $E[N_e | N_e < k_2] = \frac{1 - (1 - p_2) \cdot (k_2 \cdot e^{-\lambda_e \cdot R_v} + 1)}{p_2 \cdot e^{-\lambda_e \cdot R_v}}$.

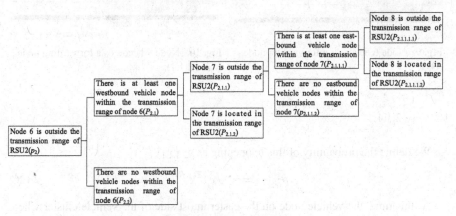

Fig. 8. Analysis of $TRSU2$

(1) Case 2.1

According to [12], we define the probability of this happening as $p_{2.1}$.

$$p_{2.1} = 1 - e^{-2\lambda_w \cdot R_v}. \tag{24}$$

When case 2.1 occurs, the easternmost vehicle node of the vehicle cluster where the westward traveling vehicle node is located in the transmission range of node 6 (i.e., node 7) is the forwarding node of the accident notification information. When node 7 is located in the transmission range of RSU2, the accident notification information will be forwarded from node 7 to RSU2. At this time, the delay caused by the direct forwarding of information is much smaller than that of storage carrying forwarding, which can be ignored. Therefore, it is only necessary to study the case where node 7 is outside the RSU2 transmission range (i.e., Case 2.1.1), as shown in Fig. 10. Denote the probability of occurrence of case 2.1.1 as $p_{2.1.1}$, according to [12], we have

$$p_{2.1.1} = 1 - (1 - e^{-\lambda_w \cdot R_v})^{k_{2.1.1}}, \tag{25}$$

where $k_{2.1.1} = 2 \cdot (d - x - E_2 + R_v)/E[d_v^w]$.

For this case, Wang et al. [5, 12–14] did not continue processing, but selected node 6 to continue carrying the accident notification information to RSU2. We propose a secondary communication method between vehicle clusters. The method first determines whether there is a vehicle node traveling eastward located in the transmission range of node 7. If it exists, the accident notification information is transmitted to the easternmost vehicle node of the vehicle cluster where the node is located, so that it transmits the accident notification information to RSU2. If it does not exist, the selected node 6 will continue to carry the accident notification information to the RSU2.

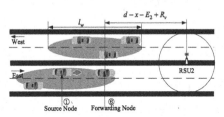
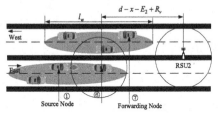

Fig. 9. Node 6 becomes a forwarding node (Case 2)

Fig. 10. Node 7 becomes a forwarding node (Case 2.1.1)

1) Case 2.1.1.1

We define the probability of this happening as $p_{2.1.1.1}$.

$$p_{2.1.1.1} = 1 - e^{-2\lambda_e \cdot R_v}. \tag{26}$$

At this time, the vehicle node on the easternmost side of the vehicle cluster where the eastbound vehicle node located in the transmission range of node 7 is located (i.e., node 8) is the forwarding node of the accident notification information. When node 8 is located in the transmission range of RSU2, the accident notification information will be forwarded from node 8 to RSU2. At this time, the delay caused by the direct forwarding of information is much smaller than that of storage carrying forwarding, which can be ignored. Therefore, it is only necessary to study the case where node 8 is outside the RSU2 transmission range (i.e., Case 2.1.1.1.1), as shown in Fig. 11. This occurs when the length of the eastbound vehicle cluster where node 8 is located satisfies the following conditions:

$$l_e < d - x - E_2 - E[d_c^e] + R_V + 2R_v.$$

With the Eq. (13), we define the probability of this happening as $p_{2.1.1.1.1}$.

$$p_{2.1.1.1.1} = 1 - (1 - e^{-\lambda_e \cdot R_v})^{k_{2.1.1.1.1}}, \tag{27}$$

where $k_{2.1.1.1.1} = (d - x - E_2 - E[d_c^e] + 3R_v)/E[d_v^e]$.

We define the conditional expectation E_3 when cluster length l_e satisfies the condition $l_e - R_v < d - x - E_2 - E[d_c^e] - R_u$.

$$\begin{aligned}
E_3 &= E[l_e | l_e < d - x - E_2 - E[d_c^e] - R_u + 3R_v] \\
&= E[d_v^e] \cdot E[N_e | N_e < k_{2.1.1.1.1}], \tag{28}
\end{aligned}$$

where

$$\begin{aligned}
E[N_e | N_e < k_{2.1.1.1.1}] &= \sum N_e \cdot f_e(N_e | N_e < k_{2.1.1.1.1}) \\
&= \sum_{N_e=1}^{k_{2.1.1.1.1}} N_e \cdot \frac{f_e(N_e)}{\Pr\{N_e < k_{2.1.1.1.1}\}} \\
&= \frac{1 - (1 - p_{2.1.1.1.1}) \cdot (k_{2.1.1.1.1} \cdot e^{-\lambda_e R_v} + 1)}{p_{2.1.1.1.1} \cdot e^{-\lambda_e R_v}}
\end{aligned}$$

In this case, the delivery delay required to deliver the accident notification information to the RSU2 can be calculated by:

$$T_{2.1.1.1.1} = (d - x - E_2 - E_3 - E[d_c^e] - R_u + 3R_v)/v_e. \tag{29}$$

2) Case 2.1.1.2

In case 2.1.1.2, the accident notification information needs to be stored and carried by node 6 and passed to the RSU2. We define the probability of this happening as $p_{2.1.1.2}$.

$$p_{2.1.1.2} = 1 - p_{2.1.1.1}. \tag{30}$$

At this time, the accident notification information delivery delay is approximated by:

$$T_{2.1.1.2} = \frac{d - x - R_u - E_2 + R_v}{v_e}. \tag{31}$$

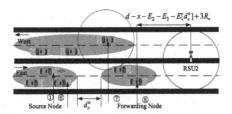

Fig. 11. Node 8 becomes a forwarding node (Case 2.1.1.1.1)

(2) Case 2.2

According to [12], we define the probability of this happening as $p_{2.2}$.

$$p_{2.2} = 1 - p_{2.1}. \tag{32}$$

At this time, node 6 will continue to store and carry accident notification information until the accident notification information is transmitted to the RSU2. The calculation expression of the delay $T_{2.2}$ required for the transmission of the accident notification information is the same as that of $T_{2.1.1.2}$.

In summary, we have

$$T_{RSU2} = p_2 \cdot [p_{2.1} \cdot p_{2.1.1}(\cdot p_{2.1.1.1} \cdot p_{2.1.1.1.1} \cdot T_{2.1.1.1.1} + p_{2.1.1.2} \cdot T_{2.1.1.2}) + p_{2.2} \cdot T_{2.2}]. \tag{33}$$

5 Numerical Analysis

To validate the accuracy and validity of the theoretical analysis model derived in the previous section, we use MATLAB to calculate and analyze the numerical results. Parameter settings are shown in Table 2.

Figure 12 shows the effects of different deployment distances of RSU and different speeds of vehicle nodes on the transmission delay of accident notification information. It can be concluded from Fig. 12: First, under the same RSU deployment distance, with the increase of the average speed of the vehicle node, the delivery delay of the accident notification information decreases. The reason is: that as the average speed of the vehicle node increases, the forwarding node can forward the accident notification information to the RSU in a shorter time. Secondly, with the increase of RSU deployment distance, the delay of accident notification information transmission increases under the same vehicle node speed. The reason is: that as the deployment distance of the two RSUs increases, the distance between the accident vehicle and the two RSUs increases, which causes the forwarding node carrying the accident notification information to take longer to enter the transmission range of the RSUs, which in turn leads to the accident notification. The increase in information delivery delay. Finally, comparing Fig. 12(a), (b), and (c), we find that with the increase in vehicle node arrival rate, the delay of accident notification information delivery decreases. The reason is: that as the arrival rate of vehicles increases, the density of vehicle nodes on the road increases, the length of the formed vehicle cluster

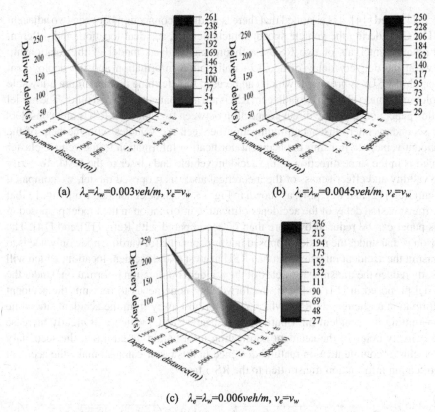

(a) $\lambda_e=\lambda_w=0.003veh/m,\ v_e=v_w$ (b) $\lambda_e=\lambda_w=0.0045veh/m,\ v_e=v_w$

(c) $\lambda_e=\lambda_w=0.006veh/m,\ v_e=v_w$

Fig. 12. Effects of different deployment distances of RSU and different speeds of vehicle nodes on the transmission delay of accident notification information

Table 2. Parameter settings.

Parameters	Values
RSU deployment distance d	5000, 7000, 9000, 11000, 13000, 15000 m
RSU transmission range R_u	250 m
Vehicle transmission range R_v	250 m
Eastbound vehicle arrival rate λ_e	0.003, 0.0045, 0.006 veh/m
Westbound vehicle arrival rate λ_w	0.003, 0.0045, 0.006 veh/m
Eastbound vehicle speed v_e	15, 20, 25, 30, 35, 40 m/s
Westbound vehicle speed v_w	15, 20, 25, 30, 35, 40 m/s
Number of iterations	1000

increases. The accident vehicle can choose a vehicle that is far away from it but closer to the RSU as a forwarding node.

In [12] and [14], it is assumed that there is no direct connection between two adjacent RSUs. Considering the vehicle speed, vehicle density, accident location, and different deployment intervals of two RSUs, a theoretical analysis model of the relationship between the delivery delay of accident notification information and the deployment interval of RSUs is proposed, which can be used to determine the maximum allowable deployment interval when two adjacent RSUs are not connected to each other. The model in this paper considers the direct connection between two adjacent RSUs and considers the secondary inter-cluster communication between the vehicles with the help of the accident vehicle to transmit the accident notification information to the vehicle cluster which is in the same direction as the accident vehicle and closer to the RSU. To verify the validity and effectiveness of the theoretical analysis proposed model, we compared it with Ref. [12] and Ref. [14], as shown in Fig. 13. It can be concluded from Fig. 13 that the transmission delay of the accident notification information in the model proposed in this paper can be reduced by more than 66% compared with Refs. [12] and [14]. The reason is that under the model proposed in this paper, the forwarding node only needs to transmit the accident information to an RSU nearest to the accident location, which will greatly reduce the transmission delay of the accident notification information. Under the model proposed in [12] and [14], the forwarding node needs to transmit the accident information to the remote RSU which is the farthest away from the accident site while transmitting the accident information to the nearest RSU, which will greatly increase the delivery delay of the accident notification information. In addition, the secondary inter-cluster communication method designed in this paper can also make the accident notification information transmitted to the RSU faster.

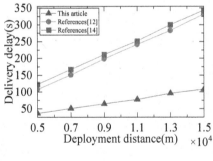

$(\lambda_e=\lambda_w=0.003veh/m, v_e=40m/s, v_w=30m/s)$

Fig. 13. Compared with references 12 and 14

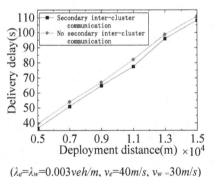

$(\lambda_e=\lambda_w=0.003veh/m, v_e=40m/s, v_w=30m/s)$

Fig. 14. Secondary inter-cluster transmission verification

To further verify the effectiveness of the secondary inter-cluster communication method designed in this paper, we design a comparative experiment considering the secondary inter-cluster transmission and not considering the secondary inter-cluster transmission, as shown in Fig. 14. It can be concluded from Fig. 14 that the delivery delay of accident notification information considering the secondary inter-cluster communication mode is significantly lower than that without considering the secondary inter-cluster transmission. The reason is that under the same conditions, the secondary transmission between clusters can transmit the accident notification information to a vehicle cluster farther away from the accident vehicle but closer to the RSU, which will significantly reduce the transmission delay of the accident notification information.

6 Conclusion

In this paper, in the highway scene with sparse vehicle nodes, RSU is deployed to improve the transmission delay of accident notification information. A theoretical analysis model is established by analyzing the relationship between the transmission delay of accident notification information and the RSU deployment distance in the VANETs environment. An inter-cluster transmission method is proposed to transmit the accident notification information to the vehicle cluster in the same driving direction and closer to the RSUs through the vehicle cluster in the opposite direction. Compared with the existing model that assumes that two adjacent RSUs are not connected and the secondary inter-cluster communication is not considered, the transmission delay of the accident notification information of our proposed model can be reduced by 66% compared with the existing model. In the future, we will implement the proposed model in the vehicular simulation environment to verify the effectiveness of our model in VANETs.

Acknowledgements. Project supported by the China University Industry-University-Research Collaborative Innovation Fund (Future Network Innovation Research and Application Project) (No. 2021FNA04017).

References

1. Bokaba, T., Doorsamy, W., Paul, B.S.: Comparative study of machine learning classifiers for modelling road traffic accidents. Appl. Sci. **12**(2), 828–846 (2022)
2. The Number of Cars Worldwide is set to double by 2040. https://www.weforum.org/agenda/2016/04/the-number-of-cars-worldwide-is-set-to-double-by-2040. Accessed 30 Aug 2022
3. Global status report on road safety 2018. https://www.who.int/publications/i/item/9789241565684. Accessed 30 Aug 2022
4. Yang, L., Ma, J.R., Zhao, X.M., Mu, K.N., Ma, J.Y.: Vehicle collision warning model based on vehicle-road collaboration. J. Highw. Transp. Res. Dev. **34**(9), 123–129 (2017)
5. Wang, Y., Zheng, J., Mitton, N.: Delivery delay analysis for roadside unit deployment in vehicular ad hoc networks with intermittent connectivity. IEEE Trans. Veh. Technol. **65**(10), 8591–8602 (2015)
6. Wu, J.L., Ye, Y.T., Wu, Y.: Roadside unit deployment algorithm based on useful contribution. J. Softw. **29**(1), 43–51 (2018)
7. Liang, Y., Wu, Z., Hu, J.: Road side unit location optimization for optimum link flow determination. Comput.-Aided Civil Infrastruct. Eng. **35**(1), 61–79 (2020)
8. Silva, C.M., Pitangui, C.G., Miguel, E.C., Santos, L.A., Torres, K.B.V.: Gamma-reload deployment: planning the communication infrastructure for serving streaming for connected vehicles. Veh. Commun. **21**, 100197 (2020)
9. Ge, J.L., Lv, W.H., Fu, S.Y., Qu, Y.X.: Research on highway roadside unit deployment based on vehicle clusters. Sci. Technol. Eng. **20**(22), 9222–9228 (2020)

10. Sou, S.I., Tonguz, O.K.: Enhancing VANET connectivity through roadside units on highways. IEEE Trans. Veh. Technol. **60**(8), 3586–3602 (2011)
11. Wisitpongphan, N., Bai, F., Mudalige, P., Sadekar, V., Tonguz, O.: Routing in sparse vehicular ad hoc wireless networks. IEEE J. Sel. Areas Commun. **25**(8), 1538–1556 (2007)
12. Wang, Y., Zheng, J., Mitton, N.: Delivery delay analysis for roadside unit deployment in intermittently connected VANETs. In: 57th IEEE Global Communications Conference, Austin, pp.155–161. IEEE (2014)
13. Liu, C., Huang, H., Du, H.: Optimal RSUs deployment with delay bound along highways in VANET. J. Comb. Optim. **33**(4), 1168–1182 (2016). https://doi.org/10.1007/s10878-016-0029-5
14. Wang, Y.D.: Modeling and an analysis of network connectivity and roadside unit deployment for vehicular ad hoc networks. Master thesis, Southeast University (2018). (in Chinese)
15. Patra, M., Mishra, S., Murthy, C.S.R.: An analytic hierarchy process based approach for optimal road side unit placement in vehicular ad hoc networks. In: 79th IEEE Vehicular Technology Conference (VTC Spring), Seoul, pp. 1–5. IEEE (2014)

CPSOCKS: Cross-Platform Privacy Overlay Adapter Based on SOCKSv5 Protocol

Gabriel-Cosmin Apostol[1], Alexandra-Elena Mocanu[1], Bogdan-Costel Mocanu[1], Dragos-Mihai Radulescu[1], and Florin Pop[1,2(✉)]

[1] Department of Computer Science, University Politehnica of Bucharest, Bucharest, Romania
{gabriel.apostol,dragos.radulescu}@stud.acs.upb.ro,
alexandra_elena.mihaita@cti.pub.ro,
{bogdan_costel.mocanu,florin.pop}@upb.ro
[2] National Institute for Research & Development in Informatics - ICI Bucharest, Bucharest, Romania
florin.pop@ici.ro

Abstract. In figures, the cybersecurity landscape is one of the most across-the-border impactable trends in the last years, especially after the begging of the COVID-19 pandemic. Therefore, by the end of Q4 of 2021, more than 281 million people have been victims of data breaches and cyber-threads, costing more than $42.96 million per day. A possible explanation is that most network operators do not provide any mechanism that blocks path tracing. Almost anybody with above-average network security knowledge can use public path tracing tools such as *traceroute*, enabling malicious users and thread factors to craft sophisticated cyber-attacks easily. Therefore, this paper proposes a cross-platform privacy overlay over the SOCKSv5 protocol. We evaluate the proposed solution in terms of latency, average throughput, and transfer rate.

Keywords: CSOCKS5 · Cloud network access patterns · Obfuscation · Privacy · Security

1 Introduction

Obfuscation of data in network transmissions represents a major research domain since it covers a large span of techniques meant to make sensitive data harder to detect. Due to internet censorship in some states, data-hiding techniques have evolved, making them less susceptible to detection. However, it is a well-known fact that there is a continuous innovation effort ongoing, both for censors and privacy advocates. One important aspect of network obfuscation is the fact that is a critical process that needs to be executed in time constraint fashion.

The information hiding domain has been researched since ancient times, and various methods have been devised since then, like the invisible ink, or the encoding of letters in music notes. More elaborate methods have been devised during

C. Yu et al. (Eds.): GPC 2022, LNCS 13744, pp. 149–161, 2023.
https://doi.org/10.1007/978-3-031-26118-3_11

the two world wars, since the army research programs have come up with many innovations, like code-words and microdots, which were microscopic texts hidden in punctuation marks. In the digital era, the increasing interest in privacy has led to a relatively large number of information-hiding methods targeting both network transmission and files.

The purpose of the research in the domain of network access patterns obfuscation is to identify novel methods meant to bypass censorship and surveillance. Since full online anonymity is not easily achievable and assumes a user must change some habits, it is important to think about people who want to achieve it. In many cases, it has been proven that they are individuals breaking laws, and these facts have led to ethical debates regarding the types of concealed activities conducted using existing technologies. On the other hand, many oppressive regimes have been exposed by whistle-blowers, which relied on traffic obfuscation technology to deliver real information worldwide.

Another positive outlook of online privacy-enhancing technologies is given by the fact that sensitive aspects of the personal life of a user can be concealed from internet service providers. The most relevant example is the access to personal health-related data, which must not be necessarily secret, but shared only between the involved parties.

It has been widely debated that internet censorship will be employed in various countries in the foreseeable future. Researchers and industry experts should continue working to craft a future in which privacy prevails against nationwide censors, providing people with non-discriminatory access to information. One of the purposes of network traffic obfuscation employment is to help circumvent abusive surveillance and censorship. Sometimes, these techniques are used even by Law Enforcement to perform sting operations without disclosing IP ranges associated with the police, as mentioned in [2].

The biggest challenge that researchers face is the lack of an adversary model. This fact hinders efforts to provide a reliable and resilient censorship bypass system and prevents researchers from realizing if a new obfuscation method is vulnerable in practice. Another research issue is the design of such systems since tradeoffs among security, performance, and ease of use must be made.

To understand restrictions imposed by the censors, researchers should have access to deep packet inspection systems employed in various countries, an impossible fact. This drawback is a speed diminishing factor for research, but at the same time, might be considered a source for various novel techniques, useful for privacy enhancement.

Another approach to developing better obfuscation methods is to understand what normal traffic means for a censor. To achieve this, scholars should be allowed to analyze large amounts of traffic captured from retention systems, but this would have an enormous impact on user data privacy.

However, a series of questions arise regarding user interaction with network obfuscation tools. It is not yet clear how users are perceiving such tools from the perspective of risks, ethics, and legal constraints. Moreover, it is not clear how people will have access to this type of software. Since the regular users could be censors as well, it is not clear how much time would take to identify, detect and defeat a new layer of obfuscation. Another open research problem is providing users with means of plausible deniability, in case they must support legal consequences in oppressing countries.

The rest of the paper is structured as follows. In the second Section, we present a detailed state-of-the-art analysis, while in the third Section, we briefly describe the SOCKSv5 protocol, on which our solution is based. Furthermore, in forth Section we present our cross-platform obfuscation solution and in the fifth Section we evaluate it. Finally, in the sixth Section, we conclude our research.

2   Related Work

To hide data in legitimate network transfers, researchers have analyzed various techniques, targeting both the timing and storage of the protocols. Any shared resource in a network can be eligible for designing a covert channel, meant to evade deep packet inspection.

Covert channels are deaplly covered as a part of the discipline called steganography, and in network environments, can be classified according to multiple criteria, depending on the way, they transfer covert data, by their reliability, by the network layer they occur or by their ability to generate additional traffic or modify the existing one.

Timing channels are created by inducing specific events at a given time and require both the receiver and the sender have internal reliable time sources. On the other hand, a storage channel is created by modifying a shared resource, directly or indirectly.

A covert timing channel involves the evaluation of specific events occurring in various protocols [9] at a certain moment. They can be achieved by manipulating the delivery times of various network layer messages, though not very reliable if the delivery route changes dynamically (load balancing, dynamic routing protocols), thus inducing unwanted noise.

In [2], the authors propose a novel real-time solution for hiding the cloud network access patterns using a chatbot implemented on the SOCKS5 protocol. An interesting use case of using chatbots in emergency situations is also presented in [15].

A covert storage channel can be achieved by modifying the fields of various protocols [10], in a manner that would not impact the transmitted data. Such examples include tunnels, which are essentially a manner of encapsulating a covert protocol in the payload data of other protocols, like HTTP or HTTPS.

Data hiding could occur in network communications starting at the network layer in the TCP/IP stack, starting with the fundamental protocols in the IP suite, including ICMP, TCP, UDP, and DNS.

ICMP (Internet Control Message Protocol) has been proposed as a good candidate, and the most common software using this technique is called Loki [4]. This software can generate ICMP packets and control their payload (Fig. 1), to encapsulate other protocols. However, this covert channel can be easily detected by employing DPI (deep packet inspection), by applying firewall rules, or by employing an IDS (intrusion detection system). Also, a traffic analysis captured on a larger period would yield a spike in the number of ICMP requests.

ICMP packet

	Bit 0 - 7	Bit 8 - 15	Bit 16 - 31
IP Header (20 bytes)	Version/IHL	Type of service	Length
	Identification		flags and offset
	Time To Live (TTL)	Protocol	Checksum
	Source IP address		
	Destination IP address		
ICMP Payload (8+ bytes)	Type of message	Code	Checksum
	Quench		
	Data (optional)		

Fig. 1. ICMP packet structure.

DNS (domain name system) tunneling has been around for more than a decade [8] and can be conducted by manipulating the name of the desired domain in the payload of the request message (). This technique can be detected by DPI systems [20], and the analysis of a capture file could provide valuable insight because the domain names are encoded using the base32 scheme.

IP steganography [11,19] assumes the manipulation of the IP header, to deliver data covertly. The eligible field proposed for this technique is called Flags whitch is used to signal a value of zero or one at a certain offset, called DF (Do not fragment). This method has a drawback since the sender and receiver must know the size of the MTU (maximum transport unit) in advance.

The TCP/IPv4 protocol could be misused [1,3] to ensure a reliable covert channel, by altering the padding bits (PAD), the urgent pointer, the reserved bits, or even by using port numbers as an alphabet (Fig. 2).

Another approach in the field of network communication steganography is the repurposing of existing layer four protocols. Therefore, many software implementations allow the cloaking of various protocols using HTTP, HTTPS, SSH, FTP, RTP, RTCP, SIP, and UDP. HTTP and HTTPS can be easily manipulated by the initiator by altering both headers and payloads. There are many methods of manipulation, starting from low-noise techniques, like changing the order of the headers to entirely modifying the request body or parameters. The detection methods rely on statistical traffic attributes [14], since these protocols are designed for great flexibility. SSH tunneling is aimed to provide two layers of protection, by employing encryption and acting as a proxy, without revealing the actual destination endpoints. Usually, SSH outbound connections are denied

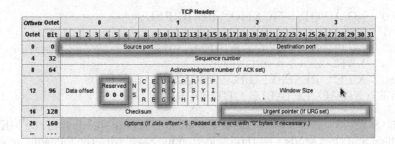

Fig. 2. TCP header structure.

due to these factors. The detection of suspicious traffic has been implemented using GMM (Gaussian Mixture Model), which, under reasonable assumptions, can be used to predict which application is being tunneled [7].

Another interesting approach ca be the usage of Peer-to-Peer tools like TOR [6], I2P [21], FreeNet [5], and UPB developed overlays like: SPIDER [16], HoneyComb [18] and AFT [17].

3   SOCKS v5 Protocol

SOCKSv5 protocol, defined in the RFC-1928 [13], relays on the compilation and the linking of a client application to the appropriate encapsulation routines of the SOCKS library. Since the latest version of this protocol, both TCP and UDP-based applications are supported. Also, this version provides strong authentication methods and addressing schemes to include IPv6 address and the domain name.

When a TCP-based client initiates a connection to a resource that is reachable only via a firewall, the application opens a TCP connection to the appropriate port on the SOCKS server system. By default, the SOCKS service is accessible on port 1080, as mentioned in [12]. If the connection request is successful, the client starts a negotiation process for the authentication method which will be further used, and then sends a relay request. After that, the SOCKS server evaluates the request, and accept or deny the connection. When the TCP-based client application connects to the SOCKS server, it sends a message which contains the following parameters, as presented in [12]:

```
VER : 1 byte;
NMETHODS : 1 byte;
METHODS : 1 - 255 bytes.
```

The VER value is set with the X'05' value to indicate the latest version of the SOCKS protocol. The NMETHODS field specifies the number of identifiers contained in the METHODS field.

The values defined, in the RFC file [12], for the METHOD field are:

- X'00' : NO AUTHENTICATION REQUIRED;
- X'01' : GSSAPI;
- X'02' : AUTHENTICATION SCHEME using username and password;
- X'03' - X'7F' : IANA ASSIGNED;
- X'80' - X'FE' : RESERVED FOR PRIVATE METHODS;
- X'FF' : NO ACCEPTABLE METHODS.

The SOCKS server selects one of the given methods specified in the METH-ODS field and then sends a method selection message, with the following scheme:

```
VER : 1 byte;
METHODS : 1 byte.
```

In the METHOD is X'FF', it means that none of the methods selected by the client are acceptable and therefore the client closes the TCP connection. Once the method negotiation process has been completed, the client sends the request details. If this phase, if the agreed method includes confidentiality and integrity validation, these requests are performed using encapsulation-dependent methods.

The request packet is formatted as follows:

```
VER : 1 byte;
CMD : 1 byte;
RSV : 1 byte;
ATYP : 1 byte;
DST.ADDR : variable;
DST.PORT : 2 bytes.
```

where:

- VER : Protocol version (X'05');
- CMD : The selected command, which can be one of the following values:
  - CONNECT : X'01';
  - BIND : X'02';
  - UDP ASSOCIATE : X'03'.
- RSV : Reserved field : X'00';
- ATYP : Address type:
  - IPv4 address : X'01';
  - Domain name address : X'03';
  - IPv6 address : X'04'.
- DST.ADDR : Destination IP address or domain name;
- DST.PORT : Destination port;

Moreover, the SOCKS server evaluates the request using the source and destination addresses and answeres with one or more messages. The message transmitted by the server is formed as follows:

```
VER : 1 byte;
REP : 1 byte;
RSV : 1 byte;
ATYP : 1 byte;
BND.ADDR : variable;
BND.PORT : 2 bytes.
```

where

- VER - Protocol version : X'05';
- The reply field, which can be one of the following values:
  - Success : X'00';
  - General SOCKS server failure : X'01';
  - Connection not permitted : X'02';
  - Network unreachable : X'03';
  - Host unreachable : X'04';
  - Connection refused : X'05';
  - TTL expired : X'06';
  - Command not supported : X'07';
  - Address not supported : X'08';
  - Unassigned value : X'09' - X'FF'.
- RSV : Reserved field : X'00';
- ATYP : Address type:
  - IPv4 address : X'01';
  - Domain name : X'03';
  - IPv6 address : X'04'.
- BND.ADDR : Bound server address;
- BND.PORT : Bound server port.

If the negotiated method in the authentication process includes integrity and confidentiality validations, the replies will use the encapsulation-dependent methods.

In the address field (BND.ADDR, DST.ADDR), the ATYP value specifies the type of address:

- X'01' : an IPv4 address (32 bits);
- X'03' : a fully-qualified domain name. The first byte contains the size of the name, followed by the value itself;
- X'04' : the address is an IP v6 address, with a length of 16 octets.

In reply to a CONNECT command sent by a client, the BND. The PORT field contains the port number for the SOCKS server and the BND.ADDR field contains its associated IP address. Usually, the BND.ADDR value is different from the IP address used, by the client, to reach the server, since these servers are often located in multiple places.

4   Proposed Solution

Our implementation aims to provide easy integration with existing applications, by combining a local SOCKS5 proxy with various legitimate protocols which allow input manipulations. Such protocols are used mostly by text chat services. These protocols will further carry data toward the final endpoint, mediated by various service providers. The endpoint shall be able to listen to incoming messages and decode them accordingly. After the endpoint decodes the requests, it will perform connections in the name of the client and deliver the data through the same protocols used by the request.

4.1   Use Case

We developed a SOCKSv5 proxy to route existing applications' network requests through a privacy overlay designed to hide network access patterns. To describe the interactions between the actors and the system, we represented a part of the functionality by including a use case diagram.

The use case starts when the actor opens the local privacy overlay application as shown in Fig. 3. Firstly, he must choose from the user interface one of the available IPv4 addresses, which will be used to listen for SOCKv5 requests.

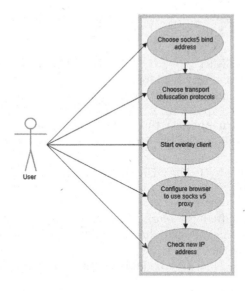

Fig. 3. CPSOCKS use case diagram.

Next, the actor must choose the transport obfuscation protocols. This method is not implemented yet, so the server simply acts like a regular SOCKSv5 proxy server.

After the bind address has been selected, the user can start the local SOCKSv5 server, allowing proxy-aware applications to use it.

To make use of our software implementation, the actor must also configure the desired applications with the address and port of the proxy server.

The last step is the verification process when the user can check the exit IP address if the application allows it. This statement is specifically true when it comes to web browsers since there are websites that allow the user to view his IP address.

4.2 Software Architecture

In the current development iteration, we managed to implement and benchmark a cross-platform SOCKS5 proxy server as shown in Fig. 4, which will be used to capture application-generated data. Our proxy server uses a limited set of SOCKS5 instructions. Currently, it supports only the TCP CONNECT method and provides no authentication method.

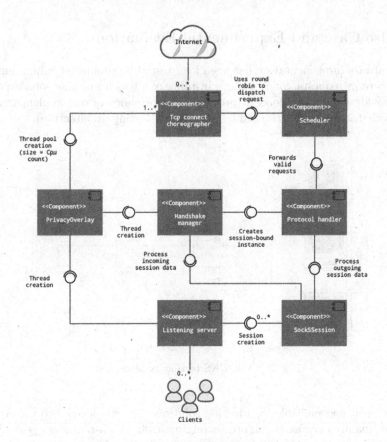

Fig. 4. CPSOCKS software architecture.

The language we chose for our implementation was Java, using the Java Development Kit version 17, mainly due to its cross-platform nature. We decided to use Java.NIO package to implement a scalable server, with a fixed number of threads and non-blocking input/output operations.

The listening server is running in its own dedicated thread, accepting connections on a specific port, enqueuing the incoming data, and dequeuing outgoing data using session-bound queues. The handshake manager runs in a separate thread, polling session queues. If any data has been pushed by the server, it employs a session-specific protocol handler to evaluate the requests and writes back its resolution afterward.

To improve speed, we decided to use multiple TCP connect choreographers, each one running in its own thread. When a connect request is dispatched, a round-robin scheduling strategy is employed to equally distribute the connection requests. The number of threads is variable, depending on the CPU cores count.

A TCP choreographer is responsible for the creation and management of multiple connections. It also routes the data generated by the actual requests and the listening server.

5   Use Case and Experimental Evaluation

Our software implementation has been load-tested by simulating a high number of concurrent requests. Since a single instance of a bench-marking solution could not provide a sufficient load to prove the performance of our implementation, we decide to employ distributed testing (Fig. 5 Testing architecture).

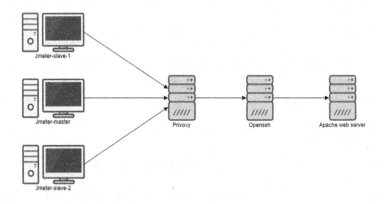

Fig. 5. CPSOCKS testing architecture.

We identified multiple load-testing solutions, but we decided to use Apache JMeter, mainly because it can orchestrate multiple bench-marking agents. However, the only downside this software implementation has is given by the fact that it supports only HTTP proxies. As a workaround, we decided to use a translation layer capable to convert HTTP proxy requests into SOCKS5 requests by using Privoxy, a software tool allowing such operations.

The benchmark we conducted had the purpose to compare our software implementation with OpenSSH, a multi-purpose application that can act as a SOCKSv5 proxy server.

The testing scenario has yielded an overall lower performance, in terms of latency, throughput, and transfer rate as shown in Fig. 6. This is happening because we are handling data transfer with additional round trips from kernel to user space. This design is acceptable because the data shall be further transformed before delivery through alternative transport channels. Also, the proxy will be accessed by a single user having a limited set of applications.

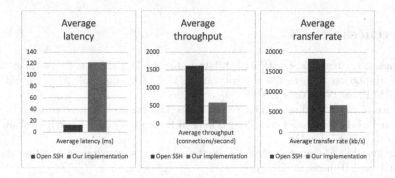

Fig. 6. CPSOCKS benchmark results.

6   Conclusions

Overlay networking, also known as software-defined networking, is a vastly-employed network-enhancing technique used to define new protocols over existing ones. The main advantage of overlay networks is that they can be defined using software, opening new opportunities in the field of security research and advancements. The most valuable enhancements they can offer are customizable encapsulations and routing protocols.

Our solution aims to provide a platform-independent method of integration with software-defined networks by leveraging the existing SOCKS-5 protocol. Even if the idea is not new, we needed a software solution that had no dependencies on external libraries and a relatively small, maintainable, and easy-to-understand code base.

The components created in this iteration are reusable and will be included in future software releases. However, they will be subject to further enhancements, mainly because now they are processing data at byte-level, rather than block-level. To achieve that, the input and output queues shall be re-engineered performance-wise.

Even if the current performance is lower than we expected, we consider that the proxy server is performing reasonably well for a single user which is not using data-intensive applications. Even though our experimental results show a reduction in performance, we still consider that our approach is an optimal solution for real-time scenarios.

Acknowledgements. The research presented in this paper was supported by project FARGO: Federated leARninG for human moBility (PN-III-P4-ID-PCE-2020-2204), the project CloudPrecis (SMIS code 2014+ 124812), project ODIN112 (PN-III-P2-2.1-SOL-2021-2-0223) and by the University Politehnica of Bucharest through the Pub-Art program. Also, this paper was partially supported by the grant POCU/993/6/13-153178, co-financed by the European Social Fund within the Sectorial Operational Program Human Capital 2014-2020, and the DECIP project (contract no. PFE57/2022).
We would also like to thank the reviewers for their valuable comments and insights.

References

1. Ahsan, K., Kundur, D.: Practical data hiding in TCP/IP. In: Proceedings of the Workshop on Multimedia Security at ACM Multimedia, vol. 2, no. 7, pp. 1–8. ACM Press, New York (2002)
2. Apostol, G.-C.: Hiding cloud network access patterns for enhanced privacy. In: 2022 21st RoEduNet Conference: Networking in Education and Research (RoEduNet), pp. 1–5. IEEE (2022)
3. Arthur, M.: Steganology and information hiding: stegop2py: embedding data in TCP and IP headers (2021)
4. Berghel, H.: Hiding data, forensics, and anti-forensics. In: Communications of the ACM, vol. 50, no. 4, pp. 15–20 (2007)
5. Clarke, I., Sandberg, O., Wiley, B., Hong, T.W.: Freenet: a distributed anonymous information storage and retrieval system. In: Federrath, H. (ed.) Designing Privacy Enhancing Technologies. LNCS, vol. 2009, pp. 46–66. Springer, Heidelberg (2001). https://doi.org/10.1007/3-540-44702-4_4
6. Dingledine, R., Mathewson, N., Syverson, P.: Tor: the second- generation onion router. Technical report, Naval Research Lab Washington DC (2004)
7. Dusi, M., Gringoli, F., Salgarelli, L.: A preliminary look at the privacy of SSH tunnels. In: 2008 Proceedings of 17th International Conference on Computer Communications and Networks, pp. 1–7. IEEE (2008)
8. Dusi, M., et al.: Detection of encrypted tunnels across network boundaries. In: 2008 IEEE International Conference on Communications, pp. 1738–1744. IEEE (2008)
9. Gray III, J.W.: Countermeasures and tradeoffs for a class of covert timing channels. Technical report (1994)
10. Kemmerer, R.A., Porras, P.A.: Covert flow trees: a visual approach to analyzing covert storage channels. In: IEEE Transactions on Software Engineering, vol. 17, no. 11, p. 1166 (1991)
11. Kundur, D., Ahsan, K.: Practical Internet steganography: data hiding in IP. In: Proceedings of the Texas WKSP. Security of Information Systems (2003)
12. Leech, M., et al.: RFC1928: SOCKS Protocol Version 5. (1996)
13. Leech, M., et al.: SOCKS protocol version 5. Technical report (1996)

14. Lin, H., Liu, G., Yan, Z.: Detection of application-layer tunnels with rules and machine learning. In: Wang, G., Feng, J., Bhuiyan, M.Z.A., Lu, R. (eds.) SpaCCS 2019. LNCS, vol. 11611, pp. 441–455. Springer, Cham (2019). https://doi.org/10.1007/978-3-030-24907-6_33

15. Mocanu, B.C., et al.: ODIN IVR-interactive solution for emergency calls handling. Appl. Sci. **12**(21), 10844 (2022)

16. Mocanu, B., et al.: SPIDER: a bio-inspired structured peer-to-peer overlay for data dissemination. In: 2015 10th International Conference on P2P, Parallel, Grid, Cloud and Internet Computing (3PGCIC), pp. 291–295. IEEE (2015)

17. Poenaru, A., Istrate, R., Pop, F.: AFT: adaptive and fault tolerant peer-to-peer overlay—A user-centric solution for data sharing. Futur. Gener. Comput. Syst. **80**, 583–595 (2018)

18. Pop, F., et al.: Trust models for efficient communication in mobile cloud computing and their applications to e-commerce. Enterp. Inf. Syst. **10**(9), 982–1000 (2016)

19. Sengupta, A., Rathor, M.: IP core steganography for protecting DSP kernels used in CE systems. IEEE Trans. Consum. Electron. **65**(4), 506–515 (2019)

20. Zander, S., Armitage, G., Branch, P.: A survey of covert channels and countermeasures in computer network protocols. IEEE Commun. Surv. Tutorials **9**(3), 44–57 (2007)

21. Zantout, B., Haraty, R., et al.: I2P data communication system. In: Proceedings of the ICN. Citeseer, pp. 401–409 (2011)

Nonlinear Optimization Method for PGC Demodulation of Interferometric Fiber-Optic Hydrophone

Jiajing Wang[✉], Zhu Kou, and Xiangtao Zhao

Dalian Naval Academy, Dalian 116018, China
601770950@qq.com

Abstract. The fiber-optic hydrophone is an advanced detection method for modern naval anti-submarine warfare and underwater weapon testing, mainly used to detect marine acoustic environments. The phase generated carrier (PGC) demodulation technology is widely applied in interferometric fiber-optic hydrophones because of its wide dynamic range, high sensitivity, and high phase measurement accuracy. An ellipse fitting optimization method based on least squares is proposed in this paper to solve the nonlinear error in the PGC demodulation process to improve the system demodulation accuracy. The experimental results show that compared with the previous demodulation, after using the least-squares ellipse fitting optimization, the relative amplitude and harmonic suppression ratio of the same frequency are greatly improved, and the demodulation accuracy is effectively enhanced.

Keywords: Fiber-optic hydrophone · PGC demodulation · Nonlinear optimization · Ellipse fitting

1 Introduction

The phase generated carrier (PGC) demodulation technology realizes the demodulation due to the phase difference caused by changing two channels of optical signals. The commonly used methods include differential cross multiplication (PGC-DCM) [1] and arctangent (PGC-Arctan) algorithm. The method of arctangent algorithm is used in our paper. The previous PGC-Arctan demodulation algorithm is still affected by the modulation depth. When the modulation depth fluctuates, the demodulation result will produce nonlinearity and cause severe harmonic distortion [2]. The traditional PGC-Arctan algorithm also does not consider the accompanying amplitude modulation and system noise caused by the internal modulation. Moreover, the two demodulated signals contain non-orthogonal nonlinear errors. The graph that is drawn by the discrete points of the two signal outputs changes from a circle to an ellipse. This paper proposes a nonlinear optimization method of ellipse fitting based on least squares as a result. By means of the nonlinear optimization algorithm of least squares, the corresponding ellipse parameters are fitted. Then the orthogonal data after nonlinear optimization is demodulated by the PGC-Arctan algorithm, which can figure out the problem of spurious amplitude modulation. By comparing the direct way of demodulation and the least squares algorithm, the effectiveness of the proposed method for nonlinear error correction in the PGC demodulation process is verified.

2 Theory

This paper adopts the Mach-Zehnder interferometric fiber-optic hydrophone. There are two PGC modulation ways, including internal and external modulations. The light intensity expression of the interference output signal channel obtained by internal modulation is as follows:

$$S = (1 + m \cos \omega_0 t)(A + B \cos[C \cos \omega_0 t + \varphi(t)]) \tag{1}$$

where T and U are constants, which are directly proportional to the power of the light source, $C \cos \omega_0 t$ is a frequency-doubled carrier signal, and $\kappa(t)$ is the phase changes caused by the changes in external physical quantities [3]. The term $(1 + m \cos \omega_0 t)$ is caused by the parasitic amplitude modulation, and m is the associated amplitude modulation index. At this time, the outputs of the two filters are:

$$S_1 = \frac{Tm}{2} + \frac{Um}{2}[J_0(C) - J_2(C)] \cos \kappa(t) - UGJ_1(C) \sin \kappa(t) \tag{2}$$

$$S_2 = \frac{Um}{2}[J_3(C) - J_1(C)] \sin \kappa(t) - UGJ_2(C) \sin \kappa(t) \tag{3}$$

In Eqs. (2) and (3), $J_i(C)$ is the Bessel's functions of the i order. G is the amplitude value. Then the output of the ratio of S_1 to S_2 is no longer linear, and the traditional PGC-Arctan demodulation scheme is not applicable. Therefore, this paper proposes an ellipse fitting algorithm based on least-squares to solve the problem of the nonlinear error caused by associated amplitude modulation. On Account of the PGC-Arctan demodulation method, two-way orthogonal signals is obtained by ellipse fitting and the least-square nonlinear optimization on the two of low-pass filtered signals. The principle block scheme of the algorithm is shown in Fig. 1.

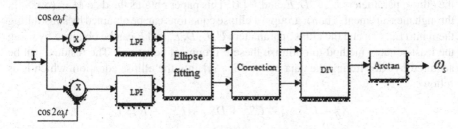

Fig. 1. The PGC demodulation method proposed in this paper.

The following part is the specific implementation of the algorithm. The final general expressions of the nonlinear error can be obtained by trigonometric transform on Eqs. (2)

and (3), which are as follows:

$$S_x = h + t \cos[\kappa(t)]$$
$$S_y = k + u \cos[\kappa(t) - \delta] \tag{4}$$

In Eq. (4), S_x and S_y are the two channels of observed signals, and the nonlinear error in the equation is calibrated by an ellipse fitting algorithm to achieve the following expression:

$$S_x = \cos[\kappa(t)]$$
$$S_y = \sin[\kappa(t)] \tag{5}$$

Equation (5) can be organized to:

$$\frac{(S_x - h)^2}{t^2} + \frac{(S_y - k)^2}{u^2} - 2\cos\delta(S_x - h)(S_y - k) \bigg/ tu = \sin^2\delta \tag{6}$$

The known standard ellipse equation is shown below:

$$x^2 + Uxy + Cy^2 + Dx + Ey + F = 0 \tag{7}$$

The parameters h, k, t, u, and δ can be obtained by comparing Eq. (6) and Eq. (7), as shown in Eq. (8):

$$\begin{cases} h = (2CD - UE) \big/ \left(U^2 - 4C\right) \\ k = (2E - UD) \big/ \left(U^2 - 4C\right) \\ t = \left[\left(h^2 + k^2C + hkU - F\right) \big/ \left(1 - U^2\big/4C\right)\right]^2 \\ u = \left(t^2\big/C\right)^{\frac{1}{2}} \\ \delta = \cos^{-1}\left[-U\big/2\sqrt{C}\right] \end{cases} \tag{8}$$

The parameters that the ellipse fitting calibration algorithm needs to calculate are the ellipse parameters U, C, D, E, and F [4]. This paper obtains the data $M = \{S_x; S_y\}_n$ through measurement. Then n groups of ellipse equations can be obtained by substituting them into Eq. (7), and the ellipse parameters U, C, D, E, and F can be obtained by using the least-squares method to optimize these n groups of equations. The residual can be acquired by substituting the data points $\{S_x; S_y\}$ into the ellipse equation, which is as follows:

$$S_{xi}^2 + US_{xi}S_{yi} + CS_{yi}^2 + DS_{xi} + ES_{yi} + F = r_i \tag{9}$$

Using the nonlinear optimization principle of the method of least-squares, the residual sum of the squares is calculated as:

$$P = \sum_{i=1}^{n} r_i^2 \tag{10}$$

To obtain the ellipse parameters that are infinitely close to the observed values, the minimum value of Eq. (10) is taken. Thus, the following equation should be satisfied:

$$\frac{\partial P}{\partial u_i}\bigg|_{i=1}^{5} = 0, \quad u_i = U, C, D, E, F \tag{11}$$

Equation (11) is expanded to:

$$
\begin{bmatrix} U \\ C \\ D \\ E \\ F \end{bmatrix} =
\begin{bmatrix}
\sum_{i=1}^{N} Sx_i^2 Iy_i^2 & \sum_{i=1}^{N} Sy_i^3 Ix_i & \sum_{i=1}^{N} Sx_i^2 Iy_i & \sum_{i=1}^{N} Sx_i Iy_i^2 & \sum_{i=1}^{N} Sx_i Iy_i \\
\sum_{i=1}^{N} Sy_i^3 Sx_i & \sum_{i=1}^{N} Sy_i^4 & \sum_{i=1}^{N} Sx_i Iy_i^2 & \sum_{i=1}^{N} Sy_i^3 & \sum_{i=1}^{N} Sy_i^2 \\
\sum_{i=1}^{N} Sx_i^2 Syi & \sum_{i=1}^{N} Sx_i Sy_i^2 & \sum_{i=1}^{N} Sx_i^2 & \sum_{i=1}^{N} Sx_i Sy_i & \sum_{i=1}^{N} Sx_i \\
\sum_{i=1}^{N} Sx_i Sy_i^2 & \sum_{i=1}^{N} Sx_i Sy_i^2 & \sum_{i=1}^{N} Sx_i Sy_i^2 & \sum_{i=1}^{N} Sy_i^2 & \sum_{i=1}^{N} Sy_i \\
\sum_{i=1}^{N} Sx_i Syi & \sum_{i=1}^{N} Sy_i^2 & \sum_{i=1}^{N} Sx_i & \sum_{i=1}^{N} Sy_i & N
\end{bmatrix}^{-1}
\begin{pmatrix} - \begin{bmatrix}
\sum_{i=1}^{N} Sx_i^3 Sy_i \\
\sum_{i=1}^{N} Sx_i^2 Sy_i^2 \\
\sum_{i=1}^{N} Sx_i^3 \\
\sum_{i=1}^{N} Sx_i^2 Sy_i \\
\sum_{i=1}^{N} Sx_i^2
\end{bmatrix} \end{pmatrix}
\tag{12}
$$

The ellipse parameters $U, C, D, E,$ and F, obtained from Eq. (12), are substituted into Eq. (8) to calculate the parameters $h, k, t, u,$ and δ. Then, by replacing the parameters $h, k, t, u,$ and δ into Eq. (4), the interference signal after calibration can be achieved as shown in Eq. (13):

$$
\begin{cases}
S_x' = (S_x - h)/t = \cos[\kappa(t)] \\
S_y' = \dfrac{[(S_y - k)/u - \cos\kappa(t)\cos\delta]}{\sin\delta} = \sin[\kappa(t)]
\end{cases}
\tag{13}
$$

where the data points $\{S'_x; I'_y\}$ are the two calibrated orthogonal interference signals.

3 Simulation

In order to analyze and verify the proposed method, relevant numerical simulations are carried out.

3.1 Simulation and Discussion of Associated Amplitude Modulation Index M

In this paper, the associated amplitude modulation index m is simulated from 0 to 0.3 by stepping of 0.05. Then, the output data before and after ellipse fitting are compared to determine whether the ellipse fitting algorithm on account of the least-squares can calibrate the nonlinear error caused by the associated amplitude modulation, and its optimization effect is verified. Figure 2 shows the variation results of the relative amplitude difference RAE, the harmonic suppression ratio HSR and the phase noise with the associated amplitude modulation index m.

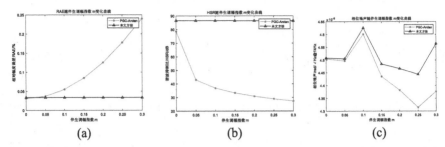

Fig. 2. Influence of the associated amplitude modulation index *m* on demodulation: (a) variation curve of RAE with m, (b) variation curve of HSR with m, and (c) variation curve of phase noise with *m*.

As shown in Fig. 2, with the increase in the parasitic amplitude modulation index, the RAE and HSR of the proposed method are basically stable at 0.03% and −87 dB, respectively, which meets the requirements of the index. However, the performance of the PGC-Arctan demodulation deteriorates sharply with the increase in the associated amplitude modulation, and the HSR even drops by 30 dB, so the demodulation function cannot be realized. In Fig. 2(c), with the increase in the parasitic amplitude modulation index m, the phase noises of the two algorithms have a slight change, both of which are less than 3×10^{-8} rad/Hz@1 kHz. However, the phase noise of the method in this paper is always higher than the PGC-Arctan demodulation, indicating that with the increase in the associated amplitude modulation index, the proposed method will increase the noise of some frequency points in the range of 1 kHz. However, this defect is negligible due to its relatively small value.

3.2 Simulation and Discussion of Two-Channel Signal Carrier Amplitude Ratio Gp/Hp

In this paper, the two-channel signal carrier amplitude ratio G_p/H_p is simulated from 1 to 10 by stepping of 1, and the output data before and after ellipse fitting are compared to determine whether the ellipse fitting algorithm on account of the least-squares can calibrate the nonlinear error caused by the amplitude deviation degree of the two-channel carriers, and its optimization effect is verified. Figure 3 shows the variation results of RAE, HSR, and phase noise with the carrier amplitude ratio G_p/H_p of the two signals.

In Fig. 3(a) and (b), with the increase in the amplitude ratio of the two signals, the RAE of the proposed technique is basically stable at 0.001%, and the HSR is stable at −87 dB, which fully meets the index requirements. However, the performance of PGC-Arctan demodulation deteriorates with the increase in the amplitude ratio of the two signals, and the demodulation function cannot be realized. As can be seen in Fig. 3(c), with the increase in the amplitude ratio of the two signals, the phase noise of the method in this paper is basically unchanged and less than 3×10^{-8} rad/Hz@1 kHz. However, the phase noise of the PGC-Arctan demodulation has always been growing with the maximum value exceeding 1.8×10^{-8} rad/Hz@1 kHz, indicating that with the increase in the amplitude ratio of the two signals, the method in this paper can effectively reduce the phase noise power and realize the optimization of nonlinear data.

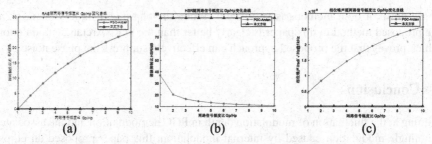

Fig. 3. Influence of two-channel signal carrier amplitude ratio G_p/H_p. On demodulation: (a) variation curve of RAE with G_p/H_p, (b) variation curve of HSR with G_p/H_p, and (c) variation curve of phase noise with G_p/H_p.

3.3 Simulation and Discussion of the Phase Difference K Between the Local Carrier and the Phase Carrier

In this paper, the phase shift K of the local carrier and the phase carrier is simulated from 0 to $\pi/4$ by stepping of $\pi/32$. The output data before and after ellipse fitting are compared to determine whether the ellipse fitting algorithm on account of the least-squares can calibrate the nonlinear error caused by the conversion of phase between the local carrier and the phase carrier, and its optimization effect is verified. Figure 4 shows the variation results of RAE, HSR, and phase noise with the phase shift K between the local carrier and the phase carrier.

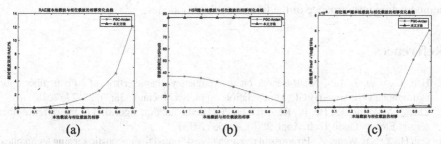

Fig. 4. Influence of the phase shift K between the local carrier and the phase carrier on demodulation: (a) variation curve of RAE with K, (b) variation curve of HSR with K, and (c) variation curve of phase noise with K.

It can be seen from Figs. 4(a) and (b) that with the increase in the conversion of phase between the local carrier and the phase carrier, the RAE of the proposed method is basically stable at 0.001%, and the total harmonic distortion is stable at -87 dB, which fully meets the requirements of the index. However, the performance of PGC-Arctan demodulation deteriorates sharply with the increase in the phase shift between the local carrier and the phase carrier, and the RAE exceeds 0.2%. Hence, the demodulation function cannot be realized. As can be seen from Fig. 4(c), with the increase of the phase shift between the local carrier and the phase carrier, the phase noise of both demodulations has some deterioration. When the phase shift is less than $5/32\pi$, the phase noise of both methods is very small. When the phase shift is less than $5/32\pi$, the proposed method in this paper is clearly better than the PGC-Arctan demodulation, which proves that the proposed approach can effectively suppress the phase noise.

4 Conclusions

Aiming at the fluctuation of modulation depth in PGC demodulation and the associated amplitude modulation caused by internal modulation, this paper proposed an ellipse fitting method based on the least-squares and the traditional PGC-Arctan demodulation method to optimize the nonlinear error. Besides, three groups of data simulations compared the proposed method with the PGC-Arctan demodulation method. The conclusions are: (1) With the increase in the associated amplitude modulation index, the method in this paper performs well in various indicators. It can effectively solve the problem of the nonlinear error caused by the associated amplitude modulation. (2) With the increase in the amplitude ratio of the two signals, the method in this paper can effectively reduce the phase noise power and realize the optimization for the nonlinear data. (3) Considering that the signal in the actual environment of the military field usually contains ocean and polarization noises, the proposed method can eliminate the influence caused by noises. (4) When the relevant parameters fluctuate, the presented method can correctly calibrate the relevant parameters with relatively good stability and effectively reduce the amplitude distortion.

In summary, the method proposed in this paper can effectively calibrate the nonorthogonal data and realize the nonlinear optimization, which helps to improve the PGC demodulation performance of the fiber-optic hydrophones.

References

1. Han, C., Cao, J., Liu, X.: Research on the frequency characteristic of LPF in fiber-optic hydrophone system with PGC demodulation. Appl. Sci. Technol. **35**(5), 23–27 (2018)
2. Zhang, A., Wang, K., He, B.: Research on PGC demodulation algorithm of interference fiber sensor. Electro-Optic Tech. Appl. **28**(6), 49–53 (2013)
3. Cai, H., Ye, Q., Wang, Z.: Progress in research of distributed fiber acoustic sensing techniques. J. Appl. Sci. **36**(1), 41–58 (2018)
4. Shi, Q.: The stability and consistency analysis of optical seismometer system using phase generated carrier in field application. In: The International Society for Optical Engineering, pp. 75081M–75081M-9. Proceedings of SPIE, Shanghai (2009)

MixKd: Mix Data Augmentation Guided Knowledge Distillation for Plant Leaf Disease Recognition

Haotian Zhang[1,2,3] and Meili Wang[1,2,3(✉)]

[1] College of Information Engineering, Northwest A&F University,
Yangling 712100, China
wml@nwsuaf.edu.cn
[2] Key Laboratory of Agricultural Internet of Things, Ministry of Agriculture,
Yangling 712100, China
[3] Shaanxi Key Laboratory of Agricultural Information Perception and Intelligent
Service, Yangling 712100, China

Abstract. Achieving fast and accurate recognition of plant leaf diseases in natural environments is crucial for plants' growth and agricultural development. The deep learning technique has been broadly used in recent years in the area of plant leaf disease classification. However, existing networks with large number of parameters are not easily deployed to farms with limited end devices and cannot be effectively utilised in natural agricultural environments. This paper proposes a data augmentation-based knowledge distillation framework for plant leaf disease recognition. We improve the traditional knowledge distillation method based on a single image by using mixed images generated from data augmentation and label annotation, significantly enhancing the recognition accuracy of the model. We have experimented on the PlantDoc dataset. The experimental results demonstrate that our approach improves recognition accuracy by up to 3.06% compared to the traditional knowledge distillation method and up to 7.23% compared to the baseline model. This study shows that the method provides a viable resolution for the diagnosis of plant foliar diseases in realistic scenarios.

Keywords: Smart agriculture · Deep learning · Disease recognition · Data augmentation · Knowledge distillation

1 Introduction

Due to environmental and bacterial factors, various diseases can appear on plant leaves, which can seriously impede plant growth [1], hinder plant industry development and ultimately lead to economic losses [2]. There are many different types

Supported by Xianyang Science and Technology Research Plan Project (2021ZDYF-NY-0014).

C. Yu et al. (Eds.): GPC 2022, LNCS 13744, pp. 169–177, 2023.
https://doi.org/10.1007/978-3-031-26118-3_13

of plant pests and diseases. The traditional way of relying on agroforestry experts to observe and identify plant pests and diseases is slow, subjective and not easily accessible. It cannot accurately and quickly identify pest, disease species or give specific countermeasures [3]. Accurate and effective identification of plant damage and delivery of targeted countermeasures has always been a pressing requirement in agroforestry, and the design of lightweight plant disease recognition models is of paramount importance. The emergence of knowledge distillation algorithms provides a new way of approaching this problem. The plant leaf, as a high-incidence area of pests and diseases, has become one of the critical resources for plant pest and disease identification because of its easy-to-obtain images.

To solve the problem of automatic recognition of plant pests and disease species, researchers have tried to use traditional machine vision and deep learning techniques to identify plant pests and disease species. Traditional machine vision-based plant pest recognition methods extract shallow features such as colour, shape and texture and then train support vector machines and BP neural networks to achieve pest and disease recognition [4,5]. However, due to factors such as unstable environment and plant leaf pest and disease area features, traditional machine vision-based plant pest and disease recognition methods are not very satisfactory in practical applications. As a result, smart sensors and deep learning technologies have been widely used in recent years in modern agriculture [6,7], generally through convolutional neural networks, to extract deep features of plant leaves for pest and disease identification. Compared to traditional machine vision methods, deep learning-based plant pest identification methods are more robust to environmental changes and have higher recognition accuracy. At the same time, constructing plant leaf pest and disease datasets is very difficult [8], as there are very few cases of pest and disease on large areas of modern farms, so data augmentation is an essential tool to enrich the dataset. This has been a great inspiration of our work.

Based on these investigations, this paper proposes a data augmentin guided knowledge distillation framework for plant leaf disease identification. Our method is a two-step process: firstly, we use a hybrid data enhancement method to generate new data from a mixture of two images and image labels. Second, we use the hybrid image labels combined with a knowledge distillation method to minimise the predictions of the teacher model and student models' predictions. We tested our approach on the PalntDoc test set. The process achieved better recognition results of plant leaf pests and diseases than the baseline model.

In summary, our contribution to this paper is threefold.

- As far as we know, We are the first to integrate data augmentation and knowledge distillation in the field of plant leaf pest and disease recognition.
- We have improved the traditional knowledge distillation method to fit the hybrid images after data augmentation.
- Compared to the benchmark model, our approach achieves a 7.23% improvement in classification accuracy on the PlantDoc dataset.

2   Relate Work

2.1   Data Augmentation

Data directly determines the upper limit of model learning. The more significant and higher the data quality, the better the model's generalisation capability. Data augmentation methods consist of two main categories, offline augmentation and online augmentation. Offline augmentation is the most performed of all conversions, essentially augmenting the size of the dataset, and it is more suitable for smaller datasets. The most common methods are cropping, flipping, rotating, scaling, shifting, Gaussian noise and colour dithering. Online augmentation is the augmentation of small batches of data (mini-batch) when feeding the data into a deep learning model. It is more suitable for large data sets and is discussed here. In contrast to offline enhancement methods, These approaches generate a new figure by mixing multiple images and generating the corresponding labels. DeVries et al. [9] cutted out random regions of the sample and filled them with 0 pixel data so that the classification result remains unchanged; Yun et al. [10] cut out some regions, but instead of filling them with 0 pixels, they randomly filled them with pixel values from the rest of the data in the dataset. As a result, the recongnition results were proportionally distributed. Zhang et al. [11] mixed two random samples, and the labels were also proportionally mixed. These give us great insight.

2.2   Knowledge Distillation

The emergence of knowledge distillation has opened up new ideas for deploying high-quality models on limited storage space as a typical type of model compression and acceleration, effectively extracting knowledge from large teacher models and imparting it to smaller student models. The concept of knowledge distillation was first introduced by Hinton et al. [12], where a more extensive network of teachers was used to instruct a smaller network of students. Knowledge transfer was done through the teacher's soft labels, and the concept of distillation temperature was introduced to smooth out the soft brands. Park et al. [13] proposed structural relationships for transferring outputs during distillation, rather than individual outputs perse, and suggested losses based on distance direction and angular direction. Yuan et al. [14] revealed the relationship between knowledge distillation and label smoothing regularization, and proposed a teacherless knowledge distillation model.

3   Method

In today's smart agricultural production, the size of hardware storage for smart devices is always a significant limiting factor, so designing lightweight models is crucial. Knowledge distillation offers a new solution to this problem. In the following paper, we will first provide some of the representations used in this paper.

We then briefly describe Mixup, a data augmentation method based on hybrid techniques, and traditional knowledge distillation methods. Next, we present our proposed MixKd approach in detail. The experimental framework is shown in Fig. 1.

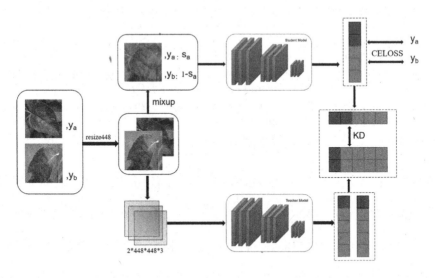

Fig. 1. The framework of our approach

Notations

We use the below notation in this paper. The original training data set $\{(I_i, y_i) \mid i \in [0, 1, ..., N-1]\}$, where $I_i \in R^{3 \times W \times H}$ and y_i refers to the input image and the label individually. Given a data pair $((I_a, y_a), (I_b, y_b))$ and hyperparameter α,The hybrid-based approach first draws a random value φ from a beta distribution Beta (α, α). Then they produce a new image $\tilde{I}$ and two label weights ρ_a and ρ_b according to φ. Here, ρ_a and ρ_b correspond to the label y_a and y_b respective.

Mixup

Blend the images and use a linear combination to combine the labels with the expression

$$\tilde{I} = \varphi \times I_a + (1 - \varphi) \times I_b \tag{1}$$

$$\rho_a = \varphi, \rho_b = 1 - \varphi \tag{2}$$

The main difference between this method and traditional data enhancement methods is how the images are blended. Mixup is a linear combination of images that can effectively improve the robustness of the neural network.

Knowledge Distillation

Knowledge distillation is a concept of dark knowledge extraction that can be understood in terms of migration learning and model compression. The focus is on proposing soft targets to complement hard targets. Soft targets are derived from the larger model's predicted output, and hard targets are labels for image categories. Soft targets contain more information about the relationships between different classes, while hard targets contain very little information, which is a one-hot form of labelling.

Loss function is expressed as

$$L = aL^{(soft)} + (1 - a) L^{(hard)} \tag{3}$$

where soft loss refers to the loss obtained from the student model's softmax output and the teacher model's softmax output, and hard loss refers to the loss obtained from the softmax output of the student model and the original labels.

MixKd for Plant Disease Recognition

One of the major limitations in the accuracy of plant leaf pest recognition is the lack of large and high-quality datasets for such tasks. Data augmentation based on blending techniques have become an essential step toward solving this problem. The method proposed in this paper combines data augmentation techniques with knowledge distillation for the first time. We first blend the two images and obtain the weights of the two original images as a percentage of the blended image and then use the weights to complete the knowledge distillation among the original and blended images. In the context of the following, we will describe our method in further detail from the perspective of soft loss.

$$L^{(soft)} = \rho_a \times L_a^{(soft)} + \rho_b \times L_b^{(soft)} \tag{4}$$

As shown in Eq. 2, Where ρ_a, ρ_b are the weights of the two images as a percentage of the mixed image. $L^{(soft)}$ refers to the output of the softmax of the blended pictures in the student model and the output of the softmax of the single picture in the teacher model for the loss.

4   Experiment

4.1   Datasets

The PlantDoc dataset [15] is a dataset for the recognition of plant leaf diseases. The dataset has 2,598 images of 13 plant species and up to 17 disease categories, with 80% of the dataset devoted to training and the remaining 20% to testing. The lack of non-laboratory datasets of sufficient size remains the main challenge in implementing vision-based plant disease recognition.

4.2   Experiment Setup

Backbone Networks and Baselines. In order to compare our method extensively with other methods, we used three network backbones as baselines in the performance comparison. In here, where not stated, we refers to the baseline as a pre-trained neural network model on the Imagenet dataset. The network constructs used include Resnet-18, 34, 50. Here we adapted their implementations in the TorchVision package for our experiments.

Data Augmentation and Knowledge Distillation Methods. We compared our approach to traditional knowledge distillation methods and one data enhancement method, namely Kd and Mixup. As previous work has not formally reported results for low-resolution fine-grained classification datasets, we implemented these results based on published code and experimented with them on the PlantDoc dataset.

Implementation Details. All deep learning models used in this study were implemented in the Pytorch deep learning framework. The experiments were performed on an 18.04.6-Ubuntu server using an Intel Xeon Platinum 8160T@2.1 GHz. An NVIDIA RTX A5000 GPU accelerated it with 24 GB of RAM.

The configuration parameters are listed in Table 1.

Table 1. Hardware and software environment.

Configuration item	Value
CPU	Inter Xeon Plantinum 8160T @2.1 GHz
GPU	NVIDIA RTX A5000 24 G
Memory	128 G
Operating system	Ubuntu 18.04.6 LTS (64-bit)
Deep learning framework	Pytorch

Training Details. In our experiments, we choose SGD as our optimizer to perform gradient descent on the model parameters, our initial learning rate is set to 0.01, and we train a total of 200 rounds, and the learning rate decreases by a factor of 10 every 80 rounds.

4.3   Experimental Results on PlantDoc Dataset

In this section, we show the results of our framework compares with the other approach. We compared MixKd with other knowledge distillation methods and data enhancement methods. We provided our proposed method's best. Comparison with data augmentation and knowledge distillation We show the results of the performance comparison in Table 2. Table 2 show the best accuracy and improvement over the baseline for each method. First, we are able to observe that our presented approach, MixKd, performs consistently outperforming its counterpart on the PlantDoc dataset. We could further find that the existing approach has limited improvement on the Resnet-34. We can further find that the improvement of the existing methods to Resnet-34 is limited. This may be mainly due to the fact that recognition by features extracted from the Resnet-34 network is insensitive to data augmentation as well as knowledge distillation. In addition, the efficacy of these approaches is more dependent on the depth of the network. As an example, Kd has achieved remarkable results on the PlantDoc dataset using just the Resnet-18 and Resnet-50 networks. while using Resnet-34, Kd even experienced performance degradation compared to Mixup. In contrast, MixKd significantly improved the baseline regardless of network depth.

We chose Resnet-101 as our teacher network. The results show that the recognition accuracy of baseline-MixKd outcompeted all other methods. Using resent18 as our baseline model, our method outperformed the baseline-Mixup method by 2.82% and the baseline-Kd method by 2.31%. Using resent34 as our baseline model, our approach outperformed baseline-Mixup method by 3.26% and baseline-Kd method by 3.32%. Using resent50 as the baseline model, our method outperformed the baseline-Mixup method by 4.18% and the baseline-Kd method by 3.85%. Our presented method achieve the highest recognition accuracy of the four methods being compared and works best in the Resnet-50 model.

Table 2. Result on plantDoc dataset

Method	Resnet-18	Resnet-34	Resnet-50
baseline	65.25%	67.79%	70.76%
baseline+Mixup	68.22%(+2.97)	69.17%(+1.38)	72.88%(+2.12)
baseline+Kd	68.73%(+3.48)	69.11%(+1.32)	73.21%(+2.45)
baseline+MixKd	**71.04%(+5.79)**	**72.43%(+4.64)**	**77.06%(+6.3)**

We also analyzed the accuracy of MixKd for each disease category, and the comparison showed that the classification accuracy was higher for leaves with prominent disease spots, such as rusted leaves. The classification accuracy was low for leaves without prominent disease spots and with leaf shape changes. For example, mosaic virus. This situation may occur because the neural network focuses on extracting leaf surface features while ignoring leaf shape when extracting features, which can also be shown in our visualization results.

4.4   Visualization Comparison

We generated a partial picture attention mechanism map using the last layer of the Resnet-50 model, showing the results that were correctly predicted by MixKd but incorrectly classified by Kd. Some background patterns distract Kd's attention, which may lead to incorrect predictions. Compared with traditional knowledge extraction methods, our proposed method can better focus on the diseases on the leaves, which is why we can obtain higher recognition accuracy (Fig. 2).

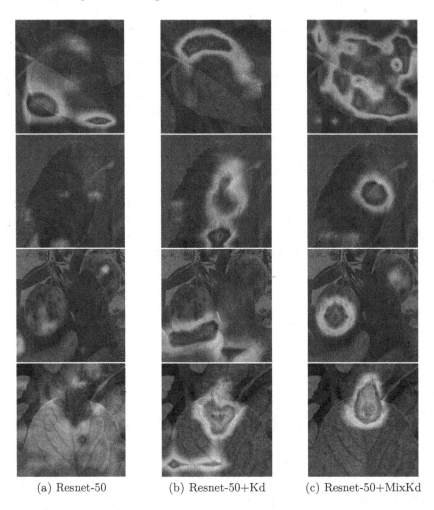

(a) Resnet-50   (b) Resnet-50+Kd   (c) Resnet-50+MixKd

Fig. 2. Visualization results of plantDoc testset

5   Conclusion

This paper presents a MixKd model for plant leaf pest identification, consisting of data augmentation and knowledge refinement. We use a teacher-student parameter to implement the distillation sub-network. In the experiments, the teacher model uses the Resnet-101 model, while the student model uses the Resnet-50, Resnet-34 and Resnet-18 models. The proposed methodology improves recognition accuracy compared to traditional knowledge extraction methods. We evaluated the model on the PlantDoc dataset and the experimental outcomes demonstrate the validity of our approach.

Acknowledgments. This research was partially funded by Xianyang Science and Technology Research PlanProject (2021ZDYF-NY-O014) and Xi'an Science and Technology Plan Project (2022JH-JSYF-O270). All supports and assistance are sincerely appreciated.

References

1. Arunnehru, J., Vidhyasagar, B.S., Anwar Basha, H.: Plant leaf diseases recognition using convolutional neural network and transfer learning. In: Bindhu, V., Chen, J., Tavares, J.M.R.S. (eds.) International Conference on Communication, Computing and Electronics Systems. LNEE, vol. 637, pp. 221–229. Springer, Singapore (2020). https://doi.org/10.1007/978-981-15-2612-1_21
2. Bhat, A., Wani, M.H., Bhat, G.M., Qadir, A., Qureshi, I., Ganaie, S.A.: Health cost and economic loss due to excessive pesticide use in apple growing region of Jammu and Kashmir. J. Appl. Hortic. **22**(3), 220–225 (2020)
3. Sankaran, S., Mishra, A., Ehsani, R., Davis, C.: A review of advanced techniques for detecting plant diseases. Comput. Electron. Agric. **72**(1), 1–13 (2010)
4. Barbedo, J.G.A.: An automatic method to detect and measure leaf disease symptoms using digital image processing. Plant Dis. **98**(12), 1709–1716 (2014)
5. Barbedo, J.G.A.: A new automatic method for disease symptom segmentation in digital photographs of plant leaves. Eur. J. Plant Pathol. **147**(2), 349–364 (2017). https://doi.org/10.1007/s10658-016-1007-6
6. Bansal, P., Kumar, R., Kumar, S.: Disease detection in apple leaves using deep convolutional neural network. Agriculture **11**(7), 617 (2021)
7. Khan, A.I., Quadri, S.M.K., Banday, S.: Deep learning for apple diseases: classification and identification. arXiv preprint arXiv:2007.02980 (2020)
8. Arivazhagan, S., Shebiah, R.N., Ananthi, S., Varthini, S.V.: Detection of unhealthy region of plant leaves and classification of plant leaf diseases using texture features. Agric. Eng. Int. CIGR J. **15**(1), 211–217 (2013)
9. DeVries, T., Taylor, G.W.: Improved regularization of convolutional neural networks with cutout. arXiv preprint arXiv:1708.04552 (2017)
10. Yun, S., Han, D., Oh, S.J., Chun, S., Choe, J., Yoo, Y.: CutMix: regularization strategy to train strong classifiers with localizable features. In: Proceedings of the IEEE/CVF International Conference on Computer Vision, pp. 6023–6032 (2019)
11. Zhang, H., Cisse, M., Dauphin, Y.N., Lopez-Paz, D.: mixup: Beyond empirical risk minimization. arXiv preprint arXiv:1710.09412 (2017)
12. Hinton, G., Vinyals, O., Dean, J., et al.: Distilling the knowledge in a neural network. arXiv preprint arXiv:1503.02531, vol. 2, no. 7 (2015)
13. Park, W., Kim, D., Lu, Y., Cho, M.: Relational knowledge distillation. In: Proceedings of the IEEE/CVF Conference on Computer Vision and Pattern Recognition, pp. 3967–3976 (2019)
14. Yuan, L., Tay, F.E.H., Li, G., Wang, T., Feng, J.: Revisiting knowledge distillation via label smoothing regularization. In: Proceedings of the IEEE/CVF Conference on Computer Vision and Pattern Recognition, pp. 3903–3911 (2020)
15. Singh, D., Jain, N., Jain, P., Kayal, P., Kumawat, S., Batra, N.: PlantDoc: a dataset for visual plant disease detection. In: Proceedings of the 7th ACM IKDD CoDS and 25th COMAD, pp. 249–253 (2020)

Multiresolution Knowledge Distillation and Multi-level Fusion for Defect Detection

Huosheng Xie[✉] and Yan Xiao

Fuzhou University, Fuzhou, China
{xiehs,200327150}@fzu.edu.cn

Abstract. Defect detection has a wide range of applications in industry, and previous work has tended to be supervised learning, which typically requires a large number of samples. In this paper, we propose an unsupervised learning method that learns knowledge about normal images by distilling knowledge from a pre-trained expert network on ImageNet to a learner network of the same size. For a given input image, we use the differences in the features of the different layers of the expert network and learner network to detect and localize defects. We show that using comprehensive knowledge makes the differences between the two networks more apparent and that combining the differences in multi-level features can make the networks more generalizable. It's worth noting that we don't need to split the picture into patches to train, and we don't need to design the learner network additionally. Our general framework is relatively simple, yet has a good detection effect. We provide very competitive results on the MVTecAD dataset and DAGM dataset.

Keywords: Defect detection · Unsupervised learning · Knowledge distillation · Multi-level fusion

1 Introduction

During the manufacturing process of industrial products, various unavoidable defects may appear in the products, such as spots, scratches, cracks, etc. Previous defect detection methods use a supervised learning approach, which requires expensive annotation costs and a low probability of defect occurrence, which can lead to a serious imbalance in the ratio of normal to defective images in the dataset. In recent years, unsupervised defect detection methods have become increasingly popular in industry [1, 2].

Usually, in unsupervised defect detection, the defect detection problem is treated as an anomaly detection problem. During training, only normal samples (samples without defects) are used, with the aim that the network learns only the features of normal samples. In the testing phase, when the input samples contain defects, the network will output results with significant differences from the normal samples, and the identification of abnormal samples (defective samples) can be achieved by detecting the differences from the normal samples. Attention was also directed to the localization of anomaly detection, expecting pixel-level localization of defective regions in the image, which is a challenging task, but it has extraordinary significance for practical applications.

C. Yu et al. (Eds.): GPC 2022, LNCS 13744, pp. 178–191, 2023.
https://doi.org/10.1007/978-3-031-26118-3_14

Much of the existing work is mainly embodied in generative models, such as autoencoders (AE) [3–6] and generative adversarial networks (GAN) [7–10]. However, due to the powerful generalization ability of the deep autoencoder, even anomalous samples containing defects can be well reconstructed, which defeats the original purpose. The literature [6] mentions that the GAN-based approach has the following two shortcomings: non-reproducibility of the results [11, 12] and data hungriness. Recent studies [13–15] have shown that these methods do not extract the semantic features well.

Using pre-trained networks can greatly increase the training speed of the model and effectively improve the accuracy of the model [16, 17]. Salehi et al. [18] proposed MKD, they extract knowledge from multiple layers of the pre-trained source network, which can better exploit the knowledge of the source network and expand the discrepancy compared to using only the last layer of information. The loss function is the similarity of the multi-layer feature maps of the source and cloner networks, using a weighted sum of MSE and cosine similarity. Moreover, the localization uses a gradient-based interpretable method, where they consider the anomalous region to be the region that makes the sudden and large change in the value of this loss, find the back-propagation gradient of the loss, and use the gradient to find the region that causes the anomaly that increases its value. We found that this method is not effective in detecting tiny defects as well as defects in texture-based products.

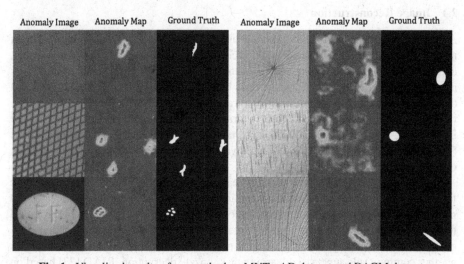

Fig. 1. Visualized results of our method on MVTecAD dataset and DAGM dataset.

To be able to better detect small defects as well as defects in textured products, we offer an alternative strategy. First, we follow the framework of knowledge distillation, distilling knowledge from one network to another. Our expert network is a pre-trained VGG16 network model [19] on the ImageNet dataset [20], and the learner network is the same size as the expert network, but the learner network is not pre-trained. In the training phase, the normal images without any defects are sent to the expert network and the learner network respectively, and the learner network will acquire different semantic information at multiple levels in the expert network, and it should not be neglected that the learner network only learns the manifold of normal data sufficiently. When an image with defects is input, the learner network and the expert network will diverge, and the greater the difference between the features of the defective and normal images, the greater their divergence will be, and the two networks will show different divergences at different layers. We only use MSE Loss to distill knowledge during the training period. In the testing phase, we use cosine similarity to obtain anomaly maps between the two networks at different levels. The value of each pixel on the anomaly maps represents the degree to which the expert network diverges from the learner network, and the more pronounced this divergence is, the more likely it is that a defect exists. By fusing multiple levels of anomaly maps, we can have excellent detection and localization effects on different types of defects (see Fig. 1). Compared with MKD [18], our method can effectively detect and locate different types of defects, especially in textured products, and has a significant improvement in the accuracy of detection. In addition to using the MVTecAD dataset, we also tested our experimental results on the DAGM dataset containing various types of texture patterns, and the data showed that our method can have excellent results in detecting defects in texture-based products.

2 Related Work

2.1 Image Reconstruction

A typical reconstruction-based approach uses an autoencoder to compress the input image. During training, the model only reconstructs the normal samples for learning, and the defective regions cannot be reconstructed well, and the presence of defects is determined based on the reconstruction error between the input data and the reconstructed data. Bergmann et al. [3] introduced structural similarity SSIM to a general autoencoder, integrating luminance, contrast and structural information to compensate for the visual inconsistency caused by reconstruction errors using the Euclidean distance metric alone. Some other methods [7–9, 21] use Generative Adversarial Network to constrain the data distribution.

Some approaches use self-supervised learning to force the model to learn semantic features of the image itself from restricted information. Golan et al. [22] subjected the normal samples to multiple geometric transformations. Similarly, Fei et al. [14] proposed that ARNet adds an attribute erasure module to the autoencoder framework to erase the color information and perform geometric transformations. Puzzle-AE [6] introduces another common self-supervised learning task, puzzle decryption. RIAD [23] based on image restoration for anomaly detection.

2.2 Feature Modeling

Yi et al. [24] improved on SVDD [25] and Deep SVDD [26] by dividing the whole image into several patches. Shi et al. [27] develop an effective feature reconstruction mechanism for anomaly detection. Cohen et al. [28] combined the idea of KNN and extraction of multilevel features to achieve good results in pixel-level localization. Wang et al. [29] achieved superior results in localization accuracy by extracting features from the ResNet intermediate layer and using a step-by-step phase multiplication method. CAVGA [30] makes clever use of the attention mechanism and expects the model to focus on the normal regions of the image. Bergmann et al. [15] first applied the knowledge distillation method to anomaly detection, since the training is based on patches, the training cost is too high and heavily depends on the size of patches.

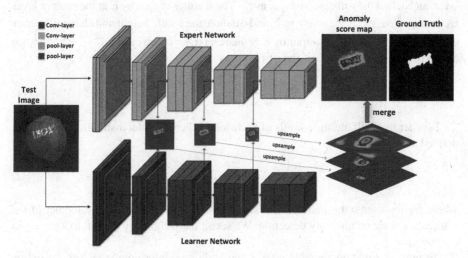

Fig. 2. Overview of our framework. During the training phase, the learner network learns the manifold of the normal data from the expert network. During the testing phase, detect and locate defects by fusing anomaly maps from multiple layers of both networks.

3 Method

This section describes in detail our proposed method for defect detection. Given a training dataset $D_{train} = \{I_1, I_2, \ldots, I_n\}$ without any defective images, we will use a pre-trained expert network E to distill knowledge to the learner network L, that detects defects in the test set, D_{test}. Once L learns the manifold of the normal data, it can assign a score to each pixel indicating how much it deviates from the training data manifold. Therefore, it has to try to learn the complete knowledge of E. The previous work related to knowledge distillation simply taught the final output information. The knowledge of the middle layer of the expert network is sometimes even better than the knowledge of the last layer [31]. For this, we encourage L to learn multi-level of knowledge, which will enable it to fully learn the normal data in the manifold. As we all know, in deep neural networks, different levels of features represent different meanings. For example, the output of the first convolutional layer is just some very simple line information, followed by possibly some shape-related information. The deeper the hierarchy goes, the easier it is to get some semantically relevant information.

Figure 2 illustrates our proposed framework.

3.1 Knowledge Distillation

In this section, we focus on how L learns the manifold of the normal data from E.

The network we use is VGG16, and the features extracted by the VGG network have demonstrated superior performance in many application directions in the field of computer vision. We call the layer where knowledge needs to be distilled the reserve layer, and define the i-th reserve layer as R_i. The features output by E at the reserve layer are called $a_E^{R_i} \in \mathbb{R}^{w \times h \times d}$, where w, h, and d denote the width, height and channel number of the feature, the features output by L at the reserve layer are called $a_L^{R_i} \in \mathbb{R}^{w \times h \times d}$, we define the distillation loss l_i of the i-th layer as

$$l_i = \frac{1}{w \times h}(a_E^{R_i} - a_L^{R_i})^2. \tag{1}$$

In order to distill multiple levels of knowledge, then the total distillation loss can be defined as

$$l = \sum_{i=1}^{N_R} \lambda_i (\frac{1}{w \times h}(a_E^{R_i} - a_L^{R_i})^2), \tag{2}$$

where N_R represents the number of reserve layers, and λ_i indicates the impact of the i-th feature scale on anomaly detection. We set all the weights by default to $\lambda_1 = \lambda_2 = \ldots = \lambda_{N_R} = 1$

To prevent some undesirable effects and additional interference factors caused by inconsistent network structures, such as inconsistent network structures leading to some differences in the output of the middle layer itself, etc. The structure of L we use is identical to that of E, the only difference being that E is pre-trained, whereas L is not. During training, we input only normal image samples and no abnormal image samples containing defects, and we keep the parameters of E unchanged while updating only the parameters of L.

The framework used for the training process is shown in Fig. 3.

3.2 Anomaly Detection

To detect possible defects contained in the images and where they are located, we feed each image into both E and L. For a normal image input without defects, the two have almost the same view of the normal image because L is well equipped with the knowledge about the normal image instilled by E. Therefore the features output by the two networks are almost identical. But for an image that contains defects, since E is pre-trained on ImageNet, and L only learns the knowledge about normal images taught by E. So the features output by the two networks for the defective image may not be consistent, and the inconsistent area is the area where the defect is located. Based on this feature, we can discern whether an image is a defective image or not, and to find the location where the defect is located.

In convolutional neural networks, local features extracted by convolutional layers are combined by subsequent convolutional layers to form more complex features. In this learning process, a hierarchy of features emerges, where lower-level convolutional layers

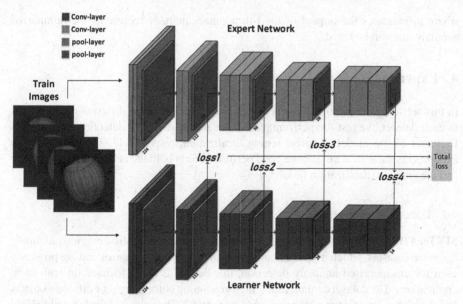

Fig. 3. The process of knowledge distillation. Knowledge from the middle and last layers of the expert network is distilled into the learner network. This knowledge is about the manifold of the normal data.

may learn lower-level features (edges, corners, etc.), while higher-level convolutional layers may learn more advanced features (dog heads, bird tails, etc.). The output of different layers of a convolutional neural network corresponds to different levels of features, and the different levels of features also represent different meanings. Intuition tells us that when an image with defects is input to the network, the features a_L^R output by each level of L will differ to a different degree from the features a_E^R output by each level of E, and we combine the differences in the features of multiple levels as a way to improve the detection accuracy of the network, and the experiments prove that our idea is correct.

In the testing phase, we use the cosine similarity to measure the difference between the features output by the two networks. The difference exhibited by two features a_L^R and a_E^R of dimension d × w × h is represented by an image of size w × h, this image is the anomaly map M_i of the current layer. M_i is formulated as

$$M_i = CosineSimilarity\left(a_E^{R_i}, a_L^{R_i}\right). \tag{3}$$

Since the size of the anomaly maps generated by each layer is inconsistent, it is necessary to upsample all the anomaly maps to a uniform size, noted as M_i^*, for subsequent fusion. M_i^* is formulated as

$$M_i^* = Upsampling\,(M_i). \tag{4}$$

The final generated anomaly map M is formulated as

$$M = \sum_1^N \beta_i M_i^*, \tag{5}$$

where β_i indicates the impact of the i-th anomaly map, N represents the number of anomaly maps to be fused.

4 Experiments

In this section, we investigate the performance of our model in different datasets, and in each dataset, we test the performance of the model in defect detection and the performance of the model in defect region localization, respectively. The results of the experiments show that we achieve good performance in both the anomaly detection task and the anomaly localization task.

4.1 Dataset

MVTecAD. MVTecAD is a dataset for anomaly detection. Unlike previous anomaly detection datasets, which mimic actual industrial production scenarios and are primarily used for unsupervised anomaly detection, this dataset is more focused on real-world applications. The dataset contains 5354 high-resolution color images of different objects (ten categories) and texture categories (five categories). The dataset is further divided into normal images for training and anomalous images for testing, with 73 anomaly types, such as scratches, dents, contamination and various structural changes, all of which are labeled at the pixel level.

DAGM2007. The DAGM2007 dataset [32] is a dataset for fabric defect detection that contains ten different classes of images. Since the DAGM2007 dataset was originally prepared for supervised and weakly supervised tasks, which contains some classification annotations, we need to discard this part of annotations and rearrange the DAGM2007 dataset to keep it consistent with the MVTec AD dataset so that it can successfully complete the task of unsupervised anomaly detection. Our reproduced DAGM2007 dataset contains 3858 images, and all ten categories are texture-based images with different sizes of defects in different categories.

4.2 Experimental Setup

Both E and L use the VGG16 network model with the same structural size. During the training phase, we choose the four final layers of each convolutional block, i.e. max-pooling layers, to be the reserve layer. Unlike the training phase, in the testing phase, for an input image of 224×224, the size of the anomaly maps output through these four storage layers are 56×56, 28×28, 14×14, and 7×7, and finally we have to upsample all the anomaly maps to the same size as the input image, which is 224×224. So for too small sizes, the upsampling process will produce some errors, which we do not want to see, so in the test phase to ensure that there is not too much interference, we discarded the output of the last storage layer 7×7, i.e. only 56×56, 28×28 and 14×14 anomaly maps are used.

Table 1. Image-level detection results on MVTecAD.

	Category	L2_AE	AnoGAN	LSA	CAVGA	VAE	MKD	PatchSVDD	STFPM	OURS
Textures	Leather	46.0	52.0	70.0	71.0	71.0	95.1	90.9	–	**99.8**
	Wood	83.0	68.0	75.0	85.0	89.0	94.3	96.5	–	**99.3**
	Carpet	67.0	49.0	74.0	73.0	67.0	79.3	92.9	–	**98.5**
	Tile	52.0	51.0	70.0	70.0	81.0	91.6	97.8	–	96.9
	Grid	69.0	51.0	54.0	75.0	83.0	78.0	94.6	–	**99.3**
Objects	Bottle	88.0	69.0	86.0	89.0	86.0	99.4	98.6	–	**99.2**
	Hazelnut	54.0	50.0	80.0	84.0	74.0	98.4	92.0	–	**98.5**
	Capsule	61.0	58.0	71.0	83.0	86.0	80.5	76.7	–	**95.8**
	Metal Nut	54.0	50.0	67.0	67.0	78.0	73.6	94.0	–	**99.6**
	Pill	60.0	62.0	85.0	88.0	80.0	82.7	86.1	–	**98.4**
	Cable	61.0	53.0	61.0	63.0	56.0	89.2	90.3	–	**92.3**
	Transistor	52.0	67.0	50.0	73.0	70.0	85.6	91.5	–	**91.8**
	Toothbrush	74.0	57.0	89.0	91.0	89.0	92.2	**100**	–	88.3
	Screw	51.0	35.0	75.0	77.0	71.0	83.3	81.3	–	**93.3**
	Zipper	80.0	59.0	88.0	87.0	67.0	93.2	**97.9**	–	97.1
	Mean	63.0	55.0	73.0	78.0	77.0	87.8	92.1	95.5	**96.5**

Table 2. Pixel-level detection results on MVTecAD.

	Category	SSIM_AE	L2_AE	AnoGAN	CNN_Dict	VAE	MKD	PatchSVDD	STFPM	OURS
Textures	Leather	78.0	75.0	64.0	59.0	92.5	98.1	97.4	**99.3**	98.6
	Wood	73.0	73.0	62.0	91.0	83.8	84.8	90.8	**97.2**	94.5
	Carpet	87.0	59.0	54.0	72.0	73.5	95.6	92.6	98.8	**99.0**
	Tile	59.0	51.0	50.0	93.0	65.4	82.8	91.4	**97.4**	96.7
	Grid	94.0	90.0	58.0	59.0	96.1	91.8	96.2	**99.0**	98.9
Objects	Bottle	93.0	86.0	86.0	78.0	92.2	96.3	98.1	**98.8**	98.5
	Hazelnut	97.0	95.0	87.0	72.0	97.6	94.6	97.5	**98.5**	98.3
	Capsule	94.0	88.0	84.0	84.0	91.7	95.9	95.8	**98.3**	91.0
	Metal Nut	89.0	86.0	76.0	82.0	90.7	86.4	98.0	**97.6**	96.5
	Pill	91.0	85.0	87.0	68.0	93.0	89.6	95.1	**97.8**	97.0
	Cable	82.0	86.0	78.0	79.0	91.0	82.4	96.8	**95.5**	94.8
	Transistor	90.0	86.0	80.0	66.0	91.9	76.5	97.0	**82.5**	80.0
	Tootbrush	92.0	93.0	90.0	77.0	98.5	96.1	98.1	98.9	**99.0**
	Screw	96.0	96.0	80.0	87.0	94.5	96.0	95.7	**98.3**	96.7
	Zipper	88.0	77.0	78.0	76.0	86.9	93.9	95.1	**98.5**	97.6
	Mean	87.0	82.0	74.0	78.0	89.3	90.7	95.7	**97.0**	95.8

For all the following experiments, we will use the framework shown in Fig. 2. In which, we do a Batch Normalization operation after each convolutional layer, not only for E but also for L. Batch Normalization allows each layer of the network to learn itself slightly more independently of the other layers. The SGD optimizer is used in the experiment, the learning rate is set to 0.3, the batch size is 32, all the input images are resized to 224 × 224, and the final output image is also 224 × 224 in size.

4.3 MVTec Anomaly Detection Dataset

As in previous work, the area under the receiver operating characteristic curve (AUROC) was used as the metric used to evaluate the experiments.

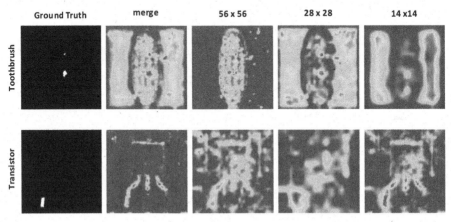

Fig. 4. Samples of bad results.

Detection. The results in Table 1 show that the multi-level feature fusion approach we used in MVTecAD has a significant improvement in detection performance compared to MKD. Especially in the texture class data, we can have good detection results regardless of the class of texture defects. However, in the Objects category, we find that the four categories of Cable, Transistor, toothbrush, and screw prevent us from going further, especially the toothbrush category, in which the detection is even worse than most of the previous methods. Observing the output anomaly maps of sizes 56, 28, and 14, as shown in Fig. 4, we found that the anomaly maps of sizes 28 and 14 judged the background region as anomalous, and after upsampling, this wrong determination was amplified, so that the accuracy of detection was greatly affected when the three were fused.

Localization. The results in Table 2 show that although our method is not optimal compared to other methods, a closer look shows that our method copes well with the various defects in the texture category. However, in the Objects category, our method does not perform well in Capsule and Transistor, especially transistor, which is less accurate than all other categories by a dozen, a very bad effect for the final average. It is easy to see that the MKD method also performs very poorly in the transistor category, so perhaps the problem arises in the VGG network itself. For the Capsule category, by observing its anomaly maps of sizes 56, 28, and 14, we find that the problem still occurs in the two small-sized anomaly maps of 28 and 14, especially the 14-sized anomaly map, which thinks that almost all the backgrounds are anomalous, and then upsampled to further expand this wrong determination result, causing the final result to become poor.

Table 3. Image-level detection results and pixel-level detection results on DAGM.

Category	Detection				Localization			
	MKD	STFPM	OURS	OURS*	MKD	STFPM	OURS	OURS*
Class1	56.3	97.8	94.9	**98.6**	56.7	**90.9**	88.7	88.1
Class2	90.6	93.8	95.7	**99.9**	97.0	94.4	97.1	**97.4**
Class3	74.6	**90.6**	57.6	75.9	83.8	**88.0**	78.7	83.3
Class4	**100**	**100**	**100**	**100**	95.4	**98.6**	97.5	97.4
Class5	68.8	83.5	95.2	**97.2**	68.9	88.9	93.2	**94.4**
Class6	90.7	**99.8**	67.6	98.3	76.1	**88.9**	75.8	80.4
Class7	49.0	**100**	98.8	**100**	71.4	**94.6**	89.5	90.6
Class8	55.0	**97.9**	82.5	97.3	75.6	**97.1**	94.5	96.7
Class9	66.0	87.5	68.8	**100**	91.1	**97.2**	79.0	94.3
Class10	94.5	**99.0**	95.6	97.8	96.8	**98.4**	98.2	**98.4**
Mean	74.6	95.0	85.7	**96.5**	81.3	**93.7**	89.2	92.1

4.4 DAGM Dataset

For the DAGM dataset, we adopt AUROC as the evaluation index used for detection as well as localization.

As shown in Table3, what can be seen is that MKD performs poorly in datasets containing complex texture class datasets and small defects, and in many of these classes, MKD does not perform well for defect detection. In contrast, our method of fusing multiple anomaly maps at different scales performs well in the datasets of these texture classes. For the default way of assigning 1/3 weight to anomaly maps of sizes 56, 28, and 14 respectively, it improves nearly 11% over MKD in terms of defect detection effect and about 8% in terms of defect localization effect. The other way of assigning 1/2, 1/3 and 1/6 weights to 56, 28 and 14 respectively, has a substantial improvement in the detection effect of defects. The weight assignment method is discussed more in Sec. 5.1. The second way of assigning weights greatly reduces the errors generated in the process of sampling to 224 on the anomaly maps of two small sizes, 28 and 14, resulting in a significant improvement of the final average effect. The table also shows that after the weight adjustment, for class6 and class9, the defect detection effect has a qualitative leap, and the defect location effect has a considerable improvement.

5   Ablation Study

5.1   Fusing Weights of Multi-scale Anomaly Maps

The outputs of our three selected convolutional blocks are 56×56, 28×28, and 14×14, respectively, with default weights of 1/3 (β) for each of the three anomaly maps, along with a set of weights of 1/2, 1/3, and 1/6 (β^*), corresponding to the layers where 56, 28, and 14 are located, respectively. Our original intention of designing the second set of weights was to worry that the size of the anomaly map output from the latter two layers was too small. In the upsampling process, because it is filling the non-existent pixel points by interpolation, it is not really detecting the presence of defects, then for these two small sizes, there is definitely an error in the upsampling process. To reduce this error, we penalize the weights of anomaly maps of small size, the smaller the anomaly map the smaller the weights assigned. In Table 4, for DAGM datasets with various complex texture classes, some defects are relatively small and some defects are not very obvious compared to normal data, the second weight assignment method avoids errors in upsampling for small anomaly maps and effectively ensures the accuracy of detection. But for the MVTecAD, many defects are relatively large, small size anomaly map in the process of upsampling by interpolation method of filling the part does not produce much error, and the integration of a variety of size anomaly map more to ensure the accuracy of the detection effect. So we use β for MVTecAD and β^* for DAGM.

Table 4. Image-level detection results and pixel-level detection results on DAGM. β is the default weight of 1/3 for each of the three layers, and β^* represents the weights of 1/2, 1/3, and 1/6.

	β		β^*	
Dataset	Detection	Localization	Detection	Localization
MVtecAD	**96.5**	95.8	90.5	**95.9**
DAGM	85.7	89.2	**96.5**	**92.1**

5.2   Number of Layers During Training

In the exception detection phase we use the output of the first three blocks of the last four blocks instead of using the output of the last block. So we thought about a question: Since we only use the first three blocks, do we need to learn about the last block? The results in Table 5 can show that even if only three blocks are used, in the learning phase, that is, the training phase, the learner network learns the complete four blocks of knowledge of the expert network, which still has some improvement for the final overall defect detection as well as localization effect.

Table 5. Ablation studies for training layers.

	3 layers		4 layers	
Dataset	Detection	Localization	Detection	Localization
MVtecAD	96.2	95.0	**96.5**	**95.8**
DAGM	85.7	89.2	**96.5**	**92.1**

5.3 Individual Layer v.s. Multi-level Layers

The experiments performed in this section are to demonstrate the necessity of our proposed fusion of multiple scale anomaly maps, and it can be clearly seen that the three different sizes of anomaly maps, 56, 28 and 14, have different detection effects for defects in different categories. For small-sized anomaly maps, the detection accuracy may be higher, but they inevitably have errors in the upsampling process, while for large-sized anomaly maps, although the detection effect is not very good, there is not much error in the upsampling process. So taking all these factors into consideration, we fused multiple scales of anomaly maps (Table 6).

Table 6. Performance with different sizes of anomaly maps.

	56	28	14	Multilevel
AUROC	87.3	89.8	88.1	**95.8**

6 Conclusion and Discussion

We show that comprehensive knowledge propagation from a pre-trained expert network to a learner network with the same structure and combining the differences in multiple intermediate layer features of the two networks are effective in detecting defects contained in images, especially for texture-like images. Our approach avoids designing learner networks and does not require the expensive training time cost based on patch.

It is worth noting that we are pre-trained on ImageNet, which may have unexpected effects if self-supervised learning is used, and that some of the network's problems present in VGG could perhaps be improved by adding some modules; nevertheless, we provide a promising direction.

Acknowledgements. This research was partially funded by Xianyang Science and Technology Research PlanProject (2021ZDYF-NY-O014) and Xi'an Science and Technology Plan Project (2022JH-JSYF-O270). All supports and assistance are sincerely appreciated

References

1. Bergmann, P, Fauser, M, Sattlegger, D, et al.: MVTec AD--A comprehensive real-world dataset for unsupervised anomaly detection. In: Proceedings of the IEEE/CVF Conference on Computer Vision and Pattern Recognition, pp. 9592–9600 (2019)
2. Mei, S., Wang, Y., Wen, G.: Automatic fabric defect detection with a multi-scale convolutional denoising autoencoder network model. Sensors **18**(4), 1064 (2018)
3. Bergmann, P., et al.: Improving unsupervised defect segmentation by applying structural similarity to autoencoders. arXiv preprint. arXiv, 1807.02011 (2018)
4. Collin, A.S, De Vleeschouwer, C.: Improved anomaly detection by training an autoencoder with skip connections on images corrupted with stain-shaped noise. In: 2020 25th International Conference on Pattern Recognition (ICPR). IEEE, pp. 7915–7922 (2021)
5. Dehaene, D, Frigo, O, Combrexelle, S, et al.: Iterative energy-based projection on a normal data manifold for anomaly localization. arXiv preprint. arXiv, 2002.03734 (2020)
6. Salehi, M., Eftekhar, A., Sadjadi N, et al.: Puzzle-ae: novelty detection in images through solving puzzles. arXiv preprint. arXiv, 2008.12959 (2020)
7. Akcay, S., Atapour-Abarghouei, A., Breckon, T.P.: Ganomaly: semi-supervised anomaly detection via adversarial training. In: Jawahar, C.V., Li, H., Mori, G., Schindler, K. (eds.) ACCV 2018. LNCS, vol. 11363, pp. 622–637. Springer, Cham (2019). https://doi.org/10.1007/978-3-030-20893-6_39
8. Schlegl, T., Seeböck, P., Waldstein, S.M., et al.: f-AnoGAN: Fast unsupervised anomaly detection with generative adversarial networks. Med. Image Anal. **54**, 30–44 (2019)
9. Perera, P., Nallapati, R., Xiang, B.: Ocgan: one-class novelty detection using gans with con-strained latent representations. In: Proceedings of the IEEE/CVF Conference on Computer Vision and Pattern Recognition, pp. 2898–2906 (2019)
10. Sabokrou, M., Khalooei, M., Fathy, M., et al.: Adversarially learned one-class classifier for novelty detection. In: Proceedings of the IEEE Conference on Computer Vision and Pattern Recognition, pp. 3379–3388 (2018)
11. Arjovsky, M., Bottou, L.: Towards principled methods for training generative adversarial networks. arXiv preprint. arXiv 1701.04862 (2017)
12. Salimans, T., Goodfellow, I., Zaremba, W., et al.: Improved techniques for training gans. In: Advances in Neural Information Processing Systems, vol. 29 (2016)
13. Wang, S., Zeng, Y., Liu, X., et al.: Effective end-to-end unsupervised outlier detection via inlier priority of discriminative network. In: Advances in Neural Information Processing Systems, vol. 32 (2019)
14. Fei, Y., Huang, C., Jinkun, C., et al.: Attribute restoration framework for anomaly detection. IEEE Trans. Multimedia **24**, 116–127 (2020)
15. Bergmann, P., Fauser, M., Sattlegger, D., et al.: Uninformed students: student-teacher anomaly detection with discriminative latent embeddings. In: Proceedings of the IEEE/CVF Conference on Computer Vision and Pattern Recognition, pp. 4183–4192 (2020)
16. Kornblith, S., Shlens, J., Le, Q.V.: Do better imagenet models transfer better?. In: Proceedings of the IEEE/CVF Conference on Computer Vision and Pattern Recognition, pp. 2661–2671 (2019)
17. Sun, R., Zhu, X., Wu, C., et al.: Not all areas are equal: transfer learning for semantic seg-mentation via hierarchical region selection. In: Proceedings of the IEEE/CVF Conference on Computer Vision and Pattern Recognition,. pp. 4360–4369 (2019)
18. Salehi, M., Sadjadi, N., Baselizadeh, S., et al.: Multiresolution knowledge distillation for anomaly detection. In: Proceedings of the IEEE/CVF Conference on Computer Vision and Pattern Recognition, pp. 14902–14912 (2021)

19. Simonyan, K., Zisserman, A.: Very deep convolutional networks for large-scale image recognition. arXiv preprint. arXiv 1409.1556 (2014)
20. Deng, J., Dong, W., Socher, R., et al.: Imagenet: a large-scale hierarchical image database. In 2009 IEEE Conference on Computer Vision and Pattern Recognition. IEEE, pp. 248–255 (2009)
21. Carrara, F., Amato, G., Brombin, L., et al.: Combining gans and autoencoders for efficient anomaly detection. In: 2020 25th International Conference on Pattern Recognition (ICPR). IEEE, pp. 3939–3946 (2021)
22. Golan, I., El-Yaniv, R.: Deep anomaly detection using geometric transformations. In: Advances in Neural Information Processing Systems, vol. 31 (2018)
23. Zavrtanik, V., Kristan, M., Skočaj, D.: Reconstruction by inpainting for visual anomaly detection. Pattern Recogn. **112**, 107706 (2021)
24. Yi, J., Yoon, S.: Patch svdd: patch-level svdd for anomaly detection and segmentation. In: Proceedings of the Asian Conference on Computer Vision (2020)
25. Tax, D.M.J., Duin, R.P.W.: Support vector data description. Machine Learn. **54**(1), 45–66 (2004). https://doi.org/10.1023/B:MACH.0000008084.60811.49
26. Ruff, L., Vandermeulen, R., Goernitz, N., et al.: Deep one-class classification. In: International Conference on Machine Learning. PMLR, pp. 4393–4402 (2018)
27. Shi, Y., Yang, J., Qi, Z.: Unsupervised anomaly segmentation via deep feature reconstruction. Neurocomputing **424**, 9–22 (2021)
28. Cohen, N., Hoshen, Y.: Sub-image anomaly detection with deep pyramid correspondences. arXiv preprint. arXiv 2005.02357 (2020)
29. Wang, G., Han, S., Ding, E., et al.: Student-teacher feature pyramid matching for unsupervised anomaly detection. arXiv preprint. arXiv 2103.04257 (2021)
30. Venkataramanan, S., Peng, K.-C., Singh, R.V., Mahalanobis, A.: Attention guided anomaly localization in images. In: Vedaldi, A., Bischof, H., Brox, T., Frahm, J.-M. (eds.) ECCV 2020. LNCS, vol. 12362, pp. 485–503. Springer, Cham (2020). https://doi.org/10.1007/978-3-030-58520-4_29
31. Zhang, R., Isola, P., Efros, A.A., et al.: The unreasonable effectiveness of deep features as a perceptual metric. In: Proceedings of the IEEE Conference on Computer Vision and Pattern Recognition, pp. 586–595 (2018)
32. Wieler, M., Hahn, T.: Weakly supervised learning for industrial optical inspection. In: DAGM Symposium In (2007)

A Novel Weighted-Distance Centralized Detection Method in Passive MIMO Radar

Tingsong Song and Jiawei Zhu[✉]

Science and Technology on Communication Information Security Control Laboratory, Jiaxing, China
zhujiaweigl@126.com

Abstract. This paper presents a weighted-distance centralized detection for passive MIMO radar (PMR) systems. In the traditional algorithm, the difficulty of target detection increases with the low SNR of target echo, because the test statistic of the data fusion center is equal weight summation and does not consider the difference of echo signal strength of targets in different paths. In the proposed method, a priori knowledge of distance can be obtained by grid detection, and the signal model is established according to the propagation and attenuation characteristics of electromagnetic waves in space. Then, based on generalized maximum likelihood ratio test (GLRT), a closed form solution of distance weighted detector is derived. Thus, centralized detection of weak targets can be realized. Finally, the simulation results show that the detection performance of the weighted-distance detection algorithm is better than that of the conventional algorithm.

Keywords: Passive MIMO radar · Centralized detection · Weighted-distance

1 Introduction

The traditional joint detection approach for passive MIMO radar systems is decentralized and treats the external radiation source and receiver in the system network as a single dual-base pair (transmitter-receiver). It detects each dual-base pair independently, with the resulting detectors fused together across pairs [1–5]. In passive dual-base processing, channel is divided into reference channel and monitoring channel. The reference channel contains direct wave signals obtained through directional antenna or digital beam forming technology. The monitoring channel contains the target echo signal, in which the direct wave component is suppressed by the reference signal. Then the cross-ambiguity function (CAF) of the received signals is computed in the distance-Doppler dimension. The targets are localized, detected and tracked by bistatic pairs in Cartesian space. Finally, final monitoring results of a certain area are obtained by data fusion according to the detection and tracking results of each bistatic pairs. The decentralized decision fusion method is widely applied in practice for its

C. Yu et al. (Eds.): GPC 2022, LNCS 13744, pp. 192–203, 2023.
https://doi.org/10.1007/978-3-031-26118-3_15

low system complexity and implementation difficulty. However, in low signal-to-noise ratio (SNR), the overall performance is poor due to the great loss of mutual information in the fusion process. Therefore, many scholars are studying the target detection problem of PMR systems and pursuing more effective target detection methods to boost the detection performance. Paper [6,7] used multiple external radiation sources and multiple passive radar receivers simultaneously to improve the target detection performance by taking advantage of spatial diversity for target detection. In the paper [8], a generalized coherent (GCC) detector is proposed to detect faint targets with a known SNR. It calculates the principal eigenvalues of the received signal correlation matrix to obtain the corresponding test statistics and then completes the target detection. Paper [9] studied the effect of DNR(direct-path signal-to-noise ratio) on the detection performance of the mutual correlation detector when the reference channel noise of passive radar is not negligible, and gave closed-form expressions for the detection probability and false alarm probability of the mutual correlation detector versus noise; Paper [10] mentioned that the mutual correlation detector degrades significantly when the DNR is low, and gave four GLRT-based tests. The paper [11] proposed a detector based on GLRT of the Gram matrix with unknown noise power, using the GLRT detector associated with all eigenvalues of the Gram matrix of the received signal, but the system contains only one noncooperative external radiation source.Paper [7] extended the single external emitter in paper [11] to multiple external emitters, and proposed a centralized fusion target detection method for PMR system based on GLRT criterion. This method makes good use of all mutual information between data for fusion, and the detection performance of the algorithm is better than that of the traditional CAF algorithm based on the Cross Ambiguity Function, but the algorithm does not use the weighting idea, in which the contribution of each receiver to the global decision is the same.

It is known that the signal strength is different between different transceiver stations, the existing methods do not take this factor into account in the above literature. Therefore, the detection method proposed in this paper divides the region of interest into a position-velocity four-dimensional grid, pre-estimates the position and velocity of the target by hypothesis testing the position of the possible targets in each grid, then assigns different weights to the signals on different propagation paths using the position a priori information, and finally makes comprehensive use of the relevant information of all observations through centralized fusion, so as to improve the detection performance of the whole system.

2   Signal Model

Consider a PMR system with M non-cooperative transmitters and K multichannel receivers. The spatial position vectors of the ith transmitter is denoted by $\mathbf{d}_i$, and the position of jth receiver by $\mathbf{r}_j$. Each receiver has N element array antennas. The position of the nth antenna of the jth receiver is defined relative to $\mathbf{r}_{j,n}$ by $\mathbf{r}_{j,n} = \mathbf{r}_j + \mathbf{\Delta}_n$, where $\mathbf{\Delta}_n$ denotes spatial offset of the nth antenna element relative to the jth receiver. Assume the sampling length of the received signal is L, the signal sampling frequency is f_s, and the sampling interval is $T_s = 1/f_s$. The emission signal of the ith transmitter is $\mathbf{x}_i = [x_i(0), x_i(T_s), \cdots, x_i((L-1)T_s)]^T \in \mathbb{C}^{L \times 1}$. By orienting the antenna to different sources or using digital beamforming methods, each receiver can obtain a direct wave signal from different sources, and these reference channels are uncorrelated. The direct wave signal from the ith transmitter received by the jth receiver can be modeled as:

$$\mathbf{y}_{ij}^d = \alpha_{ij}^d \boldsymbol{\mathcal{D}}_{ij}^d \mathbf{x}_i + \mathbf{n}_{ij}^d \tag{1}$$

where, α_{ij}^d is direct-path channel parameters, which account for various effects including, e.g., antenna gains, spreading losses and any unknown phase offsets. $\boldsymbol{\mathcal{D}}_{ij}^d = \boldsymbol{\mathcal{D}}(\tau_{ij}^d, 0)$ represents the TD variation of the direct-path signal due to the relative position and velocity of the ith transmitter and the jth receiver, where time delay is $\tau_{ij}^d = \|\mathbf{d}_i - \mathbf{r}_{j,n}\|/c$. $\mathbf{n}_{ij}^d$ is the complex Gaussian white noise of the reference channel with a mean value of 0 and a variance of σ^2.

Assume the position and the speed vector of the target are denoted by $\mathbf{s} = [x_t, y_t]^T$ and $\mathbf{v} = [\dot{x}_t, \dot{y}_t]^T$ respectively. Assuming that the equivalent effective reflected area RCS of the target in all directions is the same, and the receivers of PMR get the same power from the antenna in all directions, thus, the power attenuation β_{ti} from the ith external radiation source to the target with state $\mathbf{t} = [\mathbf{s}, \mathbf{v}]$ is only related to the relative received distance $|\mathbf{s} - \mathbf{d}_i|^2$ of the external radiation source and the target.

$$|\beta_{ti}|^2 \propto \frac{1}{|\mathbf{s} - \mathbf{d}_i|^2} \tag{2}$$

Similarly, the power attenuation β_{tj} from the target to the jth receiver with state $\mathbf{t} = [\mathbf{s}, \mathbf{v}]$ is related to the relative distance $|\mathbf{s} - \mathbf{r}_j|^2$ from the target to the receiver.

$$|\beta_{tj}|^2 \propto \frac{1}{|\mathbf{s} - \mathbf{r}_j|^2} \tag{3}$$

Then, after adding the implied distance information, the ith target echo signal collected by the jth receiver is modeled as:

$$\mathbf{y}_{ij}^t = \beta_{ti}\beta_{tj}\gamma_i \boldsymbol{\mathcal{D}}_{ij}^t \mathbf{x}_i + \mathbf{n}_{ij}^t \tag{4}$$

where γ_i is the unknown reflection coefficient of the target echo. $\boldsymbol{\mathcal{D}}_{ij}^t = \boldsymbol{\mathcal{D}}(\tau_{ij}^t, v_{ij}^t)$ represents the TD variation of the target echo signal due to the relative position and relative velocity of the ith external radiation source, the jth receiver and the target whose state is $\mathbf{t} = [\mathbf{s}, \mathbf{v}]$. The target echo time delay is $\tau_{ij}^t = (\|\mathbf{d}_i - \mathbf{s}\| + \|\mathbf{s} - \mathbf{r}_{j,n}\|)/c$. The Doppler shift is $v_{ij}^t = c\dot{\tau}_{ij}^t/\lambda$ ($\dot{\tau}_{ij}^t$ is the rate of change of the delay τ_{ij}^t). $\mathbf{n}_{ij}^t$ is the complex white Gaussian noise of the surveillance channel with mean 0 and variance σ^2.

For a given position and velocity in space, grid detection can be used to precompensate the original signal. Since the geographic distribution of the external radiation source is known information to the receiver, the TD compensation of direct wave signal is relatively simple, as follows:

$$\tilde{\mathbf{y}}_{ij}^d = \left(\mathcal{D}_{ij}^d\right)^H \mathbf{y}_{ij}^d = \alpha_{ij}^d \mathbf{x}_i + \tilde{\mathbf{n}}_{ij}^d \tag{5}$$

For the processing of the target echo signal, according to the target position and velocity $[\mathbf{u}, \dot{\mathbf{u}}]$ assumed by the grid detection, the TD variation matrix $\mathcal{D}_{ij}^t(\mathbf{u}, \dot{\mathbf{u}})$ of the target can be calculated. If the grid search result is accurate, the grid detecting that the searched position and velocity are the same as the target, $\mathbf{t} = [\mathbf{s}, \mathbf{v}] = [\mathbf{u}, \dot{\mathbf{u}}]$, then the simplified form of the compensated target echo signal is obtained:

$$\tilde{\mathbf{y}}_{ij}^t = \left(\mathcal{D}_{ij}^t\right)^H \mathbf{y}_{ij}^t = \beta_{ti}\beta_{tj}\gamma_i \mathbf{x}_i + \tilde{\mathbf{n}}_{ij}^t \tag{6}$$

When $\mathbf{t} = [\mathbf{s}, \mathbf{v}] \neq [\mathbf{u}, \dot{\mathbf{u}}]$, it is considered that only noise exists in the surveillance channel:

$$\tilde{\mathbf{y}}_{ij}^t = \tilde{\mathbf{n}}_{ij}^t \tag{7}$$

Therefore, the binary hypothesis testing model can be expressed as:

$$
\begin{aligned}
H_0 &: \begin{cases} \tilde{\mathbf{y}}_{ij}^d = \alpha_{ij}^d \mathbf{x}_i + \tilde{\mathbf{n}}_{ij}^d \\ \tilde{\mathbf{y}}_{ij}^t = \tilde{\mathbf{n}}_{ij}^t \end{cases} \\
H_1 &: \begin{cases} \tilde{\mathbf{y}}_{ij}^d = \alpha_{ij}^d \mathbf{x}_i + \tilde{\mathbf{n}}_{ij}^d \\ \tilde{\mathbf{y}}_{ij}^t = \beta_{ti}\beta_{tj}\gamma_i \mathbf{x}_i + \tilde{\mathbf{n}}_{ij}^t \end{cases}
\end{aligned} \tag{8}
$$

3   Weighted-Distance Centralized Detection Algorithm

For the above detection problems, in the PMR system environment, the direct wave, target echo and clutter signals are incident on each array. First, the original signals are preprocessed by various clutter interference suppression techniques [12–14] such as adaptive filtering to filter out useless signals. Secondly upload the preprocessed pure signal to the data fusion center, which will perform grid detection to compensate the echo signal and obtain the position prior information. Then, the compensated signal and the test statistic are weighted by the distance factor, and the different types of correlations between the observed information are used for centralized fusion processing. Finally the global test statistic is obtained so that the overall judgment can be made. The specific detection process is shown in Fig. 1.

Next, according to the test model of (8), the conditional probability density function and log-likelihood function about all observed signals $\mathbf{Y}$ under hypothesis H_0 and hypothesis H_1 can be obtained first. Then the unknown parameters can be estimated according to the GLRT criterion. Get the maximum likelihood estimation value and substitute it into the likelihood ratio test to get the specific form of the DISTANCE-GLRT tester.

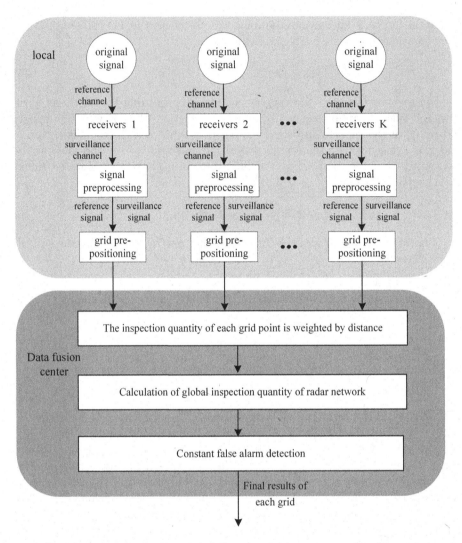

Fig. 1. Flow chart of centralized target detection in PMR system based on receiving distance weighting.

let $\boldsymbol{\alpha}_i^d = [\alpha_{i1}^d, \cdots, \alpha_{iK}^d]^T$, $\boldsymbol{\alpha}^d = \left[(\boldsymbol{\alpha}_1^d)^T, \cdots, (\boldsymbol{\alpha}_M^d)^T\right]^T$, $\boldsymbol{\gamma}_t = [\gamma_{t1}, \cdots, \gamma_{tM}]^T$.

Since the noise of the reference channel and the surveillance channel are independent between the transmission channels, the expression for the conditional probability density function (PDF) $p_1(\mathbf{Y}|\boldsymbol{\alpha}^d, \boldsymbol{\gamma}_t, \mathbf{x})$ of $\mathbf{Y}$ under the hypothesis H_1 can be expressed as:

$$p_1(\mathbf{Y}|\boldsymbol{\alpha}^d, \boldsymbol{\gamma}_t, \mathbf{x}) = \prod_{i=1}^{M} p_1^i(\mathbf{Y}_i|\boldsymbol{\alpha}_i^d, \gamma_{ti}, \mathbf{x}_i) \tag{9}$$

where $\gamma_{ti} = \beta_{ti}\gamma_i$, because both β_{ti} and γ_i are related to the ith external radiation sources, and β_{ti} is a known parameter containing the distance factor from the external radiation source to the target, so in order to facilitate the simplified calculation of the subsequent formula, the two are combined into an unknown quantity γ_{ti}.

let $c = 1/\left(\pi\sigma^2\right)^{L(M+K)}$,

$$p_1^i(\mathbf{Y}_i|\boldsymbol{\alpha}_i^d, \gamma_{ti}, \mathbf{x}_i) = c\exp\left\{-\frac{1}{\sigma^2}\sum_{j=1}^{K}\left(\left\|\tilde{\mathbf{y}}_{ij}^d - \alpha_{ij}^d\mathbf{x}_i\right\|^2 + \left\|\tilde{\mathbf{y}}_{ij}^t - \beta_{tj}\gamma_{ti}\mathbf{x}_i\right\|^2\right)\right\} \quad (10)$$

Similarly, the expression of the conditional PDF $p_1(\mathbf{Y}|\boldsymbol{\alpha}^d, \mathbf{x})$ of $\mathbf{Y}$ under the assumption of H_0 can be expressed as:

$$p_0(\mathbf{Y}|\boldsymbol{\alpha}^d, \mathbf{x}) = \prod_{i=1}^{M} p_0^i(\mathbf{Y}_i|\boldsymbol{\alpha}_i^d, \mathbf{x}_i) \quad (11)$$

where

$$p_0^i(\mathbf{Y}_i|\boldsymbol{\alpha}_i^d, \mathbf{x}_i) = c\exp\left\{-\frac{1}{\sigma^2}\sum_{j=1}^{K}\left(\left\|\tilde{\mathbf{y}}_{ij}^d - \alpha_{ij}^d\mathbf{x}_i\right\|^2 + \left\|\tilde{\mathbf{y}}_{ij}^t\right\|^2\right)\right\} \quad (12)$$

$\mathbf{Y} = \left[(\tilde{\mathbf{y}}^d)^T, (\tilde{\mathbf{y}}^t)^T\right]^T$ denotes the matrix of all signals received by the receiver including the direct and target echo signals, and $\mathbf{Y}_i = \left[(\tilde{\mathbf{y}}_i^d)^T, (\tilde{\mathbf{y}}_i^t)^T\right]^T$ denotes the matrix consisting of the signals sent to all receivers by the ith external radiation source. Notice that $\tilde{\mathbf{y}}^d = \left[(\tilde{\mathbf{y}}_1^d)^T, \cdots, (\tilde{\mathbf{y}}_M^d)^T,\right]^T$ is the direct wave signal matrix, where $\tilde{\mathbf{y}}_i^d = \left[\tilde{\mathbf{y}}_{i1}^d \cdots, \tilde{\mathbf{y}}_{iK}^d\right]^T$, and similarly, $\tilde{\mathbf{y}}^t = \left[(\tilde{\mathbf{y}}_1^t)^T, \cdots, (\tilde{\mathbf{y}}_M^t)^T\right]^T$ is the target echo matrix, where, $\tilde{\mathbf{y}}_i^t = \left[\tilde{\mathbf{y}}_{i1}^t \cdots, \tilde{\mathbf{y}}_{iK}^t\right]^T$.

Taking logarithms of (9) and (11) yields the log-likelihood functions about $\mathbf{Y}$ under H_0 and H_1, respectively, then we have:

$$\ell_0(\boldsymbol{\alpha}^d, \mathbf{x}|\mathbf{Y}) = \sum_{i=1}^{M}\ell_0^i(\boldsymbol{\alpha}_i^d, \mathbf{x}_i|\mathbf{Y}_i)$$
$$\ell_1(\boldsymbol{\alpha}^d, \gamma_t, \mathbf{x}|\mathbf{Y}) = \sum_{i=1}^{M}\ell_1^i(\boldsymbol{\alpha}_i^d, \gamma_{ti}, \mathbf{x}_i|\mathbf{Y}_i) \quad (13)$$

where the log-likelihood function about $\mathbf{Y}_i$ under the hypothesis H_0 and the hypothesis H_1 is as follows:

$$\ell_0^i(\boldsymbol{\alpha}_i^d, \mathbf{x}_i|\mathbf{Y}_i) = -\frac{1}{\sigma^2}\sum_{j=1}^{K}\left(\left\|\tilde{\mathbf{y}}_{ij}^d - \alpha_{ij}^d\mathbf{x}_i\right\|^2 + \left\|\tilde{\mathbf{y}}_{ij}^t\right\|^2\right)$$
$$\ell_1^i(\boldsymbol{\alpha}^d, \gamma_{ti}, \mathbf{x}|\mathbf{Y}) = -\frac{1}{\sigma^2}\sum_{j=1}^{K}\left(\left\|\tilde{\mathbf{y}}_{ij}^d - \alpha_{ij}^d\mathbf{x}_i\right\|^2 + \left\|\tilde{\mathbf{y}}_{ij}^t - \beta_{tj}\gamma_{ti}\mathbf{x}_i\right\|^2\right) \quad (14)$$

$\boldsymbol{\beta}_t = [\beta_{t1}, \cdots, \beta_{tK}]^T$ characterizes relative intensity of received echo, and the maximum likelihood solution of the unknown parameters is obtained according to (14):

$$\hat{\alpha}_{ij}^d = \left(\mathbf{x}_i^H \mathbf{x}_i\right)^{-1} \mathbf{x}_i^H \tilde{\boldsymbol{y}}_{ij}^d$$
$$\hat{\gamma}_{ti} = \left(\mathbf{x}_i^H \mathbf{x}_i\right)^{-1} \mathbf{x}_i^H \tilde{\boldsymbol{y}}_i^d \boldsymbol{\beta}_t \left(\boldsymbol{\beta}_t^H \boldsymbol{\beta}_t\right)^{-1} \tag{15}$$

Substituting (15) into (14) and let $\mathbb{Y}^{td} = \left\|\tilde{\boldsymbol{y}}_{ij}^d\right\|^2 + \left\|\tilde{\boldsymbol{y}}_{ij}^t\right\|^2$, then the result is given:

$$\ell_0^i(\mathbf{x}_i|\mathbf{Y}_i) = -\frac{1}{\sigma^2} \sum_{j=1}^{K} \left(\mathbb{Y}^{td} - \frac{\mathbf{x}_i^H \tilde{\boldsymbol{y}}_{ij}^d (\tilde{\boldsymbol{y}}_{ij}^d)^H \mathbf{x}_i}{\mathbf{x}_i^H \mathbf{x}_i}\right) \tag{16}$$

$$\ell_1^i(\mathbf{x}_i|\mathbf{Y}_i) = -\frac{1}{\sigma^2} \sum_{j=1}^{K} \left(\mathbb{Y}^{td} - \frac{\mathbf{x}_i^H \tilde{\boldsymbol{y}}_{ij}^d (\tilde{\boldsymbol{y}}_{ij}^d)^H \mathbf{x}_i}{\mathbf{x}_i^H \mathbf{x}_i} - \frac{\mathbf{x}_i^H (\tilde{\boldsymbol{y}}_{ij}^t \beta_{tj})(\tilde{\boldsymbol{y}}_{ij}^t \beta_{tj})^H \mathbf{x}_i}{\boldsymbol{\beta}_t^H \boldsymbol{\beta}_t \mathbf{x}_i^H \mathbf{x}_i}\right) \tag{17}$$

$\boldsymbol{\mathcal{B}}_t = diag(\beta_{t1}, \beta_{t2}, \cdots, \beta_{tK})$ characterizes the diagonal array of target echo reception relative intensity for each receiver, $\tilde{\boldsymbol{y}}_i^t$ can be converted to a new target echo matrix $\tilde{\boldsymbol{y}}_i^{t'}$ after relative intensity correction.

$$\tilde{\boldsymbol{y}}_i^{t'} = \frac{\tilde{\boldsymbol{y}}_i^t \boldsymbol{\mathcal{B}}_t}{\sqrt{\boldsymbol{\beta}_t^H \boldsymbol{\beta}_t}} \tag{18}$$

Then $\mathbf{Y}_i' = \left[(\tilde{\boldsymbol{y}}_i^d)^T, (\tilde{\boldsymbol{y}}_i^{t'})^T\right]^T$, and (17) can be further reduced to:

$$\ell_1^i(\mathbf{x}_i \mid \mathbf{Y}_i) = -\frac{1}{\sigma^2} \sum_{j=1}^{K} \left(\mathbb{Y}^{td} - \frac{\mathbf{x}_i^H \tilde{\boldsymbol{y}}_{ij}^d (\tilde{\boldsymbol{y}}_{ij}^d)^H \mathbf{x}_i}{\mathbf{x}_i^H \mathbf{x}_i} - \frac{\mathbf{x}_i^H \tilde{\boldsymbol{y}}_{ij}^{t'} (\tilde{\boldsymbol{y}}_{ij}^{t'})^H \mathbf{x}_i}{\mathbf{x}_i^H \mathbf{x}_i}\right)$$
$$= -\frac{1}{\sigma^2} \sum_{j=1}^{K} \left(\mathbb{Y}^{td} - \frac{\mathbf{x}_i^H \mathbf{Y}_i' \mathbf{Y}_i'^H \mathbf{x}_i}{\mathbf{x}_i^H \mathbf{x}_i}\right) \tag{19}$$

The log-likelihood function derived above can be viewed as a function about the unknown vector parameter $\mathbf{x}_i$. For a single external radiation source signal, The third term in (16) is consistent with the form of Rayleigh entropy:

$$y(\mathbf{x}_i) = \sum_{j=1}^{K} \left(\frac{\mathbf{x}_i^H \tilde{\boldsymbol{y}}_{ij}^d (\tilde{\boldsymbol{y}}_{ij}^d)^H \mathbf{x}_i}{\mathbf{x}_i^H \mathbf{x}_i}\right) = \frac{\mathbf{x}_i^H \tilde{\boldsymbol{y}}_i^d (\tilde{\boldsymbol{y}}_i^d)^H \mathbf{x}_i}{\mathbf{x}_i^H \mathbf{x}_i} \tag{20}$$

Denote $\lambda_1(\cdot)$ as the principal eigenvalue in the matrix and $\mathbf{v}_1(\cdot)$ as the matrix principal eigenvector. According to the nature of Rayleigh entropy, when the unknown parameter $\mathbf{x}_i = \mathbf{v}_1\left(\tilde{\boldsymbol{y}}_i^d (\tilde{\boldsymbol{y}}_i^d)^H\right)$, the function $y(\mathbf{x}_i) = \lambda_1\left(\tilde{\boldsymbol{y}}_i^d (\tilde{\boldsymbol{y}}_i^d)^H\right)$ reaches a maximum. In addition, $\lambda_1\left((\tilde{\boldsymbol{y}}_i^d)^H \tilde{\boldsymbol{y}}_i^d\right) = \lambda_1\left(\tilde{\boldsymbol{y}}_i^d (\tilde{\boldsymbol{y}}_i^d)^H\right)$, and because

$\left(\left(\tilde{\mathbf{y}}_i^d\right)^H \tilde{\mathbf{y}}_i^d\right)$ is smaller in dimension compared to $\left(\tilde{\mathbf{y}}_i^d \left(\tilde{\mathbf{y}}_i^d\right)^H\right)$, the eigenvalues are more convenient to calculate, so the maximum value of the function $y(\mathbf{x}_i)$ is:

$$\max y\left(\mathbf{x}_i\right) = \lambda_1 \left(\left(\tilde{\mathbf{y}}_i^d\right)^H \tilde{\mathbf{y}}_i^d\right) \tag{21}$$

Substituting (21) back into (16), we obtain:

$$\max \ell_0^i \left(\hat{\mathbf{x}}_i \mid \mathbf{Y}_i\right) = -\frac{1}{\sigma^2} \sum_{j=1}^{K} \left(\left\|\tilde{\mathbf{y}}_{ij}^d\right\|^2 + \left\|\tilde{\mathbf{y}}_{ij}^t\right\|^2 - \lambda_1 \left(\left(\tilde{\mathbf{y}}_i^d\right)^H \tilde{\mathbf{y}}_i^d\right)\right) \tag{22}$$

Similarly

$$\max \ell_1^i \left(\hat{\mathbf{x}}_i \mid \mathbf{Y}_i'\right) = -\frac{1}{\sigma^2} \sum_{j=1}^{K} \left(\left\|\tilde{\mathbf{y}}_{ij}^d\right\|^2 + \left\|\tilde{\mathbf{y}}_{ij}^t\right\|^2 - \lambda_1 \left(\mathbf{Y}_i'^H \mathbf{Y}_i'\right)\right) \tag{23}$$

According to the GLRT criterion, the expression of the global test statistic can be expressed as:

$$T_{pmr} = \max \ell_1 \left(\hat{\mathbf{x}} \mid \mathbf{Y}'\right) - \max \ell_0 \left(\hat{\mathbf{x}} \mid \mathbf{Y}'\right) \underset{H_0}{\overset{H_1}{\gtrless}} \eta \tag{24}$$

Combining (14), (22), (23) and (24), the expressions of the global test statistic can finally be obtained as follows.

$$T_{pmr} = \frac{1}{\sigma^2} \sum_{i=1}^{M} \alpha_{ti}^2 \left(\lambda_1 \left(\mathbf{Y}_i'^H \mathbf{Y}_i'\right) - \lambda_1 \left(\left(\tilde{\mathbf{y}}_i^d\right)^H \tilde{\mathbf{y}}_i^d\right)\right) \underset{H_0}{\overset{H_1}{\gtrless}} \eta \tag{25}$$

where η is the constant false alarm threshold, which is calculated according to the following equation:

$$\eta = Q_\zeta^{-1} \left(1 - P_{fa}\right) \tag{26}$$

4   Simulations

In order to confirm the detection algorithm designed in this paper, simulation experiments are performed and simulation results are given for the previously designed detection algorithm in this section. In the simulation, it is assumed that the external radiation source signals are independent of each other. The objects used for comparison are the conventional GLRT detector and the conventional CAF detector in the paper [7]. It is known that the SNR and the signal length of the target echo may affect the detection performance, so this part will show how the detection performance of the two algorithms varies with these parameters and verify whether the weighted-distance detection algorithm is better than the conventional algorithm through simulation experiments.

Simulation 4.1: Set the number of external radiation sources $M = 4$, the number of passive radar receivers $K = 4$, the received signal length $L = 2048$, the constant false alarm probability $P_{fa} = \{10^{-3}, 10^{-4}\}$, and the DNR -10 dB. Considering the difference in signal power attenuation caused by the distance from the target to the receiver and from the external radiation source to the receiver, set β_{tj} and β_{ti} to have four relative magnitude values $\beta_{tj} = \{0.2, 0.4, 0.6, 0.8\}$, $\beta_{ti} = \{0.2, 0.4, 0.6, 0.8\}$. Taking the SNR of the weakest target echo signal received by the receiver (the target is farther away and the receiver receives a weaker echo strength) as the reference for the horizontal coordinate variable, Fig. 2 shows the comparison of the DISTANCE-GLRT detector and the conventional detector with SNR for different false alarm probabilities and DNR.   As

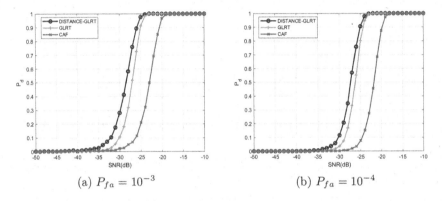

(a) $P_{fa} = 10^{-3}$ (b) $P_{fa} = 10^{-4}$

Fig. 2. Comparison of the detection probability of DISTANCE-GLRT detector and conventional detector with SNR at low DNR

the simulation results shown in Fig. 2, Under the low DNR, the DISTANCE-GLRT detection algorithm shows better detection performance compared with the traditional algorithm under the constant false alarm probability, because the algorithm takes into account the difference in the strength of the target echo signals of different paths due to the propagation distance, the greater the strength of the target echo with shorter total distance, the greater the weight of the corresponding test quantity, which improves the positive effect of the receiving base station with relatively strong received echoes on the global decision; so its detection performance is more excellent than the traditional algorithm. In addition, because the centralized fusion detection of the traditional GLRT method utilizes the mutual information amount between different surveillance channels in contrast with the traditional CAF method, the performance is somewhat better than the traditional CAF method at low DNR.

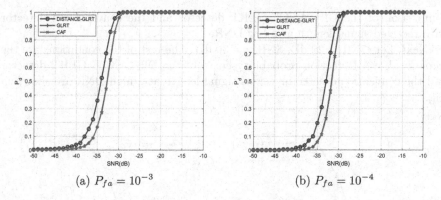

(a) $P_{fa} = 10^{-3}$ (b) $P_{fa} = 10^{-4}$

Fig. 3. Comparison of the performance of DISTANCE-GLRT detector and conventional detector with SNR at high DNR

Next, consider DNR = 20 dB, and the rest of the simulation conditions are the same as above. A comparison of the DISTANCE-GLRT detector and the conventional detector with SNR at high DNR is shown in Fig. 3.

Figure 3 show that the detection performance of the DISTANCE-GLRT algorithm is also better than the conventional algorithm at high DNR. The difference is that the detection performance of the conventional CAF detector is almost close to that of the conventional GLRT detector, which is because the gain generated when the signal is centralized in the data fusion center has limited impact on the overall performance with high SNR at reference channel.

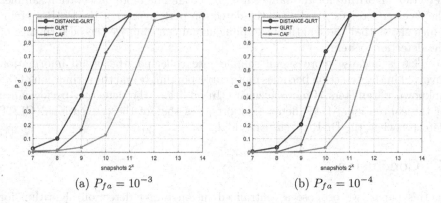

(a) $P_{fa} = 10^{-3}$ (b) $P_{fa} = 10^{-4}$

Fig. 4. Comparison of the performance of DISTANCE-GLRT detector and conventional detector with SNR at low direct wave signal-to-noise ratio

Simulation 4.2: Let DNR = -10 dB, SNR = −23 dB, and other simulation scenario, false alarm probability and reference channel SNR parameters are set to the scenario and conditions in Simulation 4.1. Figure 4 shows a performance comparison of the DISTANCE-GLRT detector and the conventional detector with received signal length at a low DNR.

Next, Let DNR = 20 dB, SNR = −30 dB. The rest of the conditions are the same as above. Detection probability curve of the DISTANCE-GLRT detector and the conventional detector versus signal length at a high DNR are shown in Fig. 5.

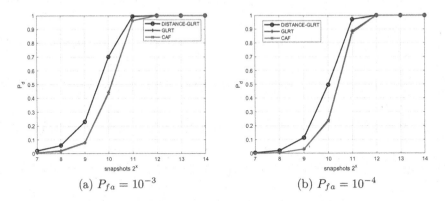

(a) $P_{fa} = 10^{-3}$ (b) $P_{fa} = 10^{-4}$

Fig. 5. Comparison of the performance of DISTANCE-GLRT detector and conventional detector with SNR at low direct wave signal-to-noise ratio

As shown in Fig. 4 and Fig. 5, the detection performance of the DISTANCE-GLRT detection algorithm is somewhat more outstanding than the conventional detection algorithm for the same sampling data. The more observation samples the receiver receives, the better the detection performance of the PMR system, which shows that the length of sampling signal can improve the performance of the detection system.

This is because more signal samples are collected, the useful information involved in the detection becomes more correspondingly and the estimation of the unknown parameters is more accurate. In addition, the detection performance of the conventional CAF detector approaches that of the conventional GLRT detector when the DNR is relatively high.

5   Conclusion

In this paper, we propose a centralized fusion target detection algorithm for PMR systems based on received distance weighting, which estimates the target's position and velocity in advance by dividing the region of interest into a position-velocity four-dimensional grid, hypothesis testing the position of the possible targets in each grid, and using the priori position information to assign different weights to signals on different propagation paths, and finally the fusion center to perform the test. The article builds a signal model based on the attenuation characteristics of electromagnetic wave propagation in space, and derives a constant false alarm detector based on the GLRT criterion. Finally, the simulation results show that the DISTANCE-GLRT method outperforms the conventional GLRT detector and the conventional CAF detector with the same DNR and sampling number.

References

1. Malanowski, M., Kulpa, K., Suchozebrski, R.: Two-stage tracking algorithm for passive radar. In: 2009 12th International Conference on Information Fusion, pp. 1800–1806. IEEE Press (2009)
2. Malanowski, M., Kulpa, K.: Two methods for target localization in multistatic passive radar. IEEE Trans. Aerosp. Electron. Syst. **48**(1), 572–580 (2012)
3. Schroeder, A., Edrich, M., Winkler, V.: Multi-illuminator passive radar performance evaluation. In: 2012 13th International Radar Symposium, pp. 61–64. IEEE Press (2012)
4. Daun, M., Nickel, U., Koch, W.: Tracking in multistatic passive radar systems using DAB/DVB-T illumination. Signal Process. **92**(6), 1365–1386 (2012)
5. Klein, M., Millet, N.: Multireceiver passive radar tracking. IEEE Aerosp. Electron. Syst. Mag. **27**(10), 26–36 (2012)
6. Hack, D.E., Patton, L.K., Himed, B., et al.: Centralized passive MIMO radar detection without direct-path reference signals. IEEE Trans. Signal Process. **62**(11), 3013–3023 (2014)
7. Hack, D.E., Patton, L.K., Himed, B., et al.: Detection in passive MIMO radar networks. IEEE Trans. Signal Process. **62**(11), 2999–3012 (2014)
8. Bialkowski, K.S., Clarkson, I.V.L., Howard, S.D.: Generalized canonical correlation for passive multistatic radar detection. In: 2011 IEEE Statistical Signal Processing Workshop (SSP), pp. 417–420. IEEE Press (2011)
9. Liu, J., Li, H., Himed, B.: On the performance of the cross-correlation detector for passive radar applications. Signal Process. **113**, 32–37 (2015)
10. Cui, G., Liu, J., Li, H., et al.: Signal detection with noisy reference for passive sensing. Signal Process. **108**, 389–399 (2015)
11. Liu, J., Li, H., Himed, B.: Two target detection algorithms for passive multistatic radar. IEEE Trans. Signal Process. **62**(22), 5930–5939 (2014)
12. Palmer, J. E., Searle, S. J.: Evaluation of adaptive filter algorithms for clutter cancellation in passive bistatic radar. In: 2012 IEEE Radar Conference, pp. 0493–0498. IEEE Press (2012)
13. Colone, F., O'hagan, D.W., Lombardo, P., et al.: A multistage processing algorithm for disturbance removal and target detection in passive bistatic radar. IEEE Trans. Aerosp. Electron. Syst. 45(2), 698–722 (2009)
14. Zemmari, R.: Reference signal extraction for GSM passive coherent location. In: 11-th InternationaL Radar Symposium, pp. 1–4. IEEE Press (2010)

Performer: A Resource Demand Forecasting Method for Data Centers

Wenkang Qi, Jieqian Yao, Jialun Li, and Weigang Wu[✉]

School of Computer Science and Engineering, Sun Yat-Sen University,
Guangzhou, China
{qiwk,yaojq6,lijlun3}@mail2.sysu.edu.cn, wuweig@mail.sysu.edu.cn

Abstract. Predicting the resource demands of online tasks plays an important role in data centers, which can help cloud providers to better allocate resources and to schedule tasks. To cope with the huge number of workloads in a data center, workloads are usually clustered first and then prediction is conducted for each cluster. However, training different models for different clusters separately reduces the overall utilization of the data in the data center, potentially reducing the prediction ability of the whole predicting system. Inspired by federated learning, we propose Performer, a Transformer-based forecasting model for clustered massive time-series. Each cluster of workloads is viewed as a local dataset owned by a training worker and all workers cooperate to train a global prediction model, while local models are trained by workers respectively. By combining global model and local models in an encoder-decoder architecture, Performer can learn global information and local information to perform predictions while keeping low model deployment costs. By splitting time-series into blocks and calculating self-attention inner-blocks, Performer keeps good prediction accuracy with lower computation cost than other Transformer-based time-series forecasting methods, making it more suitable for data center usage. Experiments on an online tasks workload dataset show Performer is an effective method in the scenario of cluster-based forecasting.

Keywords: Cloud data centers · Time-series forecasting · Transformer

1 Introduction

With the rapid development of cloud computing, there are more and more cloud data centers around the world. Most of them are using container technology, putting tasks into container, and scheduling containers to run on different servers. Forecasting the resource demands of container becomes essential for these data centers because the forecasting result is an important basis for allocating computing resources to containers and scheduling containers [1,2].

There are usually thousands of online tasks running on data centers, so it is impractical to train a model for each task, which will introduce huge model

C. Yu et al. (Eds.): GPC 2022, LNCS 13744, pp. 204–214, 2023.
https://doi.org/10.1007/978-3-031-26118-3_16

training and deploying overhead. In data centers, containers are usually clustered into several clusters, and forecasting models are equipped for each cluster [3,4]. For every container, the data center uses the model of the cluster that they belong to perform forecasting. This approach achieves a compromise between prediction accuracy and computational overhead. On the one hand, under most circumstances, models trained for clustered data get worse prediction accuracy than models trained for one single task, because the latter is entirely based on the same container. On the other hand, this approach reduces the number of forecasting models in data centers from the number of tasks to the number of clusters. Compared to the huge amount of workload records in data centers, each cluster-based model can only access a little part of them. If data centers can introduce more information to each forecasting model, the ability of data centers on resource demand forecasting can be future improved.

In federated learning [5], there are clients which have non-iid(non-independent identical distribution) data, and they try to train a global model for all clients, so each client can make the use of data from other clients. However, the global model may not fits every single client well, so people try to modify the model to adapt certain client's data to get better performance, which is called personalization learning [6]. One of the most common ideas for personalization technology is setting global and local models, using a global model to learn data from every client and a local model for the data of the specific client, the result is a combination of these two models. The clustered containers in data centers can be seen as clients that have non-iid data, too. This inspired us that a global-local architecture can be used for clustered containers in data centers.

Besides the problem of the clustered containers, time-series forecasting methods for resource demand forecasting face more challenges. Except for forecasting accuracy, data centers also want forecasting models to have low computational costs to save resources and long-term forecasting ability for better scheduling. In the time-series forecasting area, the traditional statistical methods like ARIMA [7] and VAR [8] lack the ability of forecasting complex time-series. Machine learning methods like xgboost [9] and RNN [10] can achieve accurate short-term forecasts, but for long-range forecasting, they have to use an iterative way, the gradual accumulation of errors and the slow inference speed make these methods ineffective for long term forecasting. Transformer-based models can perform accurate forecasting and are good at handling long input and output, but the high complexity of the self-attention mechanism hinders its usage in data centers.

Motivated by the above, in this paper, we propose Performer, a Transformer-based model for resource demand forecasting in data centers. Forecasting models in data centers do not require privacy protection, so Performer does not simply follow the approach in federated learning, it uses its idea and designs novel methods. To predict better for the clustered containers, Performer set global model and local models to learn global and local features separately and combines them to get the result in an encoder-decoder architecture, improving data utilization and improving overall forecast accuracy in data centers. To meet the requirements of cloud data centers on the accuracy, efficiency, and prediction length of prediction models, we use Transformer-based encoder and decoder in Performer and use the splitting method to improve computing efficiency. The contributions of this paper are summarized as follows:

- We propose a new resource demand forecasting model Performer. Considering the cluster-based forecasting approach that is widely used in data centers and inspired by federated learning, Performer set a global model to learn global data, and local models to learn data in a cluster. By combining them to predict, Performer gains better accuracy and remains low model deployment costs.
- Encoder in Performer, which is based on Transformer, uses a splitting method to split time-series into blocks and only calculate self-attention inner blocks, reducing computational cost and maintaining accuracy.
- Experiments on a clustered online task dataset show Performer outperforms other baseline methods.

2   RelatedWork

2.1   Time-Series Forecasting

Resource demand forecasting is a kind of time-series forecasting. Due to its widespread usage, various methods have been developed for it. It can be divided into three main categories. Statistical methods are popularly used because of their interpretability and simplicity. Among them, the most influential method is ARIMA [7,11], which achieves reasonable forecasting results by differencing non-stationary processes to stationary, but it only can handle univariate time-series forecasting. These methods are still being used and improved for resource forecasting. Razdca et al. [12] use a sliding window algorithm to forecast, they apply different variants of it and choose the best of them to use. Calheiros et al. [13] use ARIMA to perform workload prediction and study its impact on cloud applications. Machine learning methods often use models such as Random Forest [14] and XGBoost [9], which have good computational efficiency and interpretability. Cortez et al. [2] use random forest and XGBoost to predict CPU utilization. These years, deep learning methods achieve the most superior performance by using RNN, CNN, and their variants. Tang et al. [1] combine wavelet neural network techniques and linear regression to predict workload. Song et al. [15] use the LSTM network to predict host load. Lai et al. [16] uses recurrent-skip connections with CNN to extract local and long-range dependency patterns to perform multivariate forecasting. However, as mentioned above, these methods fail to solve long-term input and output problems, which are demanded by data center's actual needs.

After Transformer [17] is invented for language translation, it has become the most powerful model in natural language sequence modeling. These years, many works try to introduce it to time series modeling. Wu et al. [18] introduce it for disease propagation forecasting and gained better performance than RNN and CNN methods, while Song et al. [19] adapt it for clinical time series analysis.

Zerveas et al. [20] use Transformer for unsupervised representation learning of multivariate time series and tests it on regression and classification tasks. Lim et al. [21] use Transformer for multi-horizon univariate forecasting, allowing interpretation of temporal dynamics. Ma et al. [22] use Transformer to infer missing values in multivariate time and geographic coordinate sequences. To solve the high computational complexity of self-attention mechanism, LogTrans [23] and Informer [24] use some sparse variants of self-attention mechanism, reducing the computational complexity to $O(LlogL)$. To the best of our knowledge, though Transformer-based models show great potential for long-term forecasting, there is still no research that uses them for resource demand forecasting.

2.2   Personalization Techniques of Federated Learning

Federated learning [5] is proposed to enable machine learning models to learn from private decentralized data without compromising privacy. Traditional federated learning produces one shared model for all clients. However, in most cases, the clients in federated learning have non-iid data, and the shared model can not suit every client's data well. Personalization learning [6] aims to personalize the global model, making it performs better for individual clients. Many methods have been proposed for personalization learning. Mansour et al. [25] suggest cluster clients and train model for each cluster, which is similar to the approach data centers used for forecasting. Also, methods like transfer learning, multi-task learning, and meta-learning are introduced for personalize [26–28]. Combining global and local models is another popular method for personalization learning. Instead of learning a global model and fine-tuning them for individual clients, Hanzel et al. try to train global model and local models, using both of them to get the result. They also invited Loopless Local Gradient Descent(LLGD) to perform gradient descent on both global and local model while maintaining privacy.

As mentioned above, data centers have clustered clusters with non-iid data, they usually train the model on each cluster. But in the view of personalization technology, introducing global information into these models can help them to predict. On the other hand, forecasting methods in data centers do not need to concern about privacy, which means they do not need to follow the methods in federated learning.

3   Methodology

Performer is a resource demand forecasting model for the clustered containers in data centers. Here we provide a brief problem definition for time-series forecasting. Given an input $X = \{x_1^t, ..., x_{L^x}^t | x_i^t \in R^{d_x}\}$ and forecasting steps L_y, the model needs to predict the next L_y steps $Y = \{y_1^t, ..., y_{L^y}^t | y_i^t \in R^{d_y}\}$. The dimension of X and Y can be 1 for univariate forecasting or bigger for multivariate forecasting $(d_x \geq 1, d_y \geq 1)$.

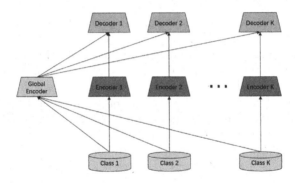

Fig. 1. Performer model architecture.

3.1 Overall Architecture

Figure 1 shows the overall architecture of Performer. As mentioned above, Performer is designed for clustered massive time-series. For K clusters in total, Performer set one global encoder, K local encoders, and K local decoders. During training, Performer uses a two-step approach. For the first step, Performer trains its encoders. The global encoder receives all data in the dataset to train itself, while the Kth local encoder only receives data in the Kth cluster. In the second step, the Kth decoder receives data from the Kth cluster, and the output of the global encoder and Kth local encoder to perform prediction. In this step, the encoders are frozen and only decoders are trained. When using Performer to predict, the data from the Kth cluster are passed through the global encoder, Kth encoder, and Kth decoder to get the result. A detailed description of the encoder and decoder can be found below.

3.2 Encoder

The encoder of Performer is a three-layer Transformer-based model, as shown in Fig. 2. It first embeds the input to the model dimension by positional embedding and 1-D convolution embedding. The 1-D convolution embedding with a filter of size 3 and stride 1 can obtain a more suitable representation of the sequence by enriching information from context and reducing the impact of anomaly points. At each stage, it splits the input into several non-overlapping blocks, then calculates self-attention only inner blocks. By splitting the time-series, the encoder can significantly reduce its computation amount by reducing the number of dot-product in self-attention calculation, and the model is focused on the short-term pattern in the time-series. Next, all blocks are put together to complete the subsequent calculation. Following the setting in Transformer, the result of self-attention is the input to a layerNorm layer and two MLP layers to get the final output. Residual connection is added after the multi-head self-attention layer and the MLP layers. On the left part of Fig. 2, we demonstrate an example when the block number is 4. After stages 1 and 2, the output is fed into the

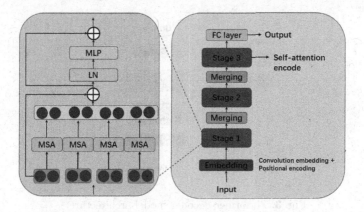

Fig. 2. Performer encoder model architecture.

merging layer. At the merging layer, it merges adjacent blocks by deforming the input and feeding it into a fully connected layer to enrich the information of its output. In our implementation, after two merging layers, the initial four blocks are merged into one, so the final self-attention operation has a respective field of the whole input, allowing the model to learn about long-term dependency in input data.

During training, the encoder's output is put into a fully connected layer to perform prediction, and MSE loss is used to propagate back and train the encoder. After training, the fully connected layer is removed, and the attention encode is used as decoder's input when training decoder or performing prediction. This decoupled, two-step training method allows the global encoder and local encoders to learn about their field first and it can be used by the decoder with a stable output.

3.3 Decoder

The decoder of Performer has a two-layer architecture, as shown in Fig. 3. For the input of the decoder, it is inputted into a self-attention layer to get an enhanced representation after embedding. The second layer is an attention layer, the Query matrix in attention calculation is from the previous layer, while the Key and Value matrix are the contact of global encoder's output and local encoder's output. The result is input into a fully connected layer to get the final result. After generating an output, the model uses MSE loss to propagate back and train itself, the global and local encoders are freezed and do not participate in training.

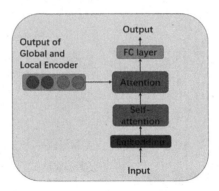

Fig. 3. Performer decoder model architecture.

4   Experiment

4.1   Dataset

The dataset is sampled from a production data center of an Internet company, including 297 containers running online tasks. Though many types of resource utilization data are recorded, we only use CPU utilization for forecasting, as CPU utilization is the most important data for task scheduling. Each time-series record contains the 7-day record of the container, with a sampling frequency of 30 s and a length of 20160. The dataset is splitted into train/valid/test set by the ratio of 5:1:1.

The 297 records of workload are clustered into 10 clusters by the most commonly used K-means algorithm. Figure 4 shows the centroid of 10 container clusters and Table 1 shows the number of each cluster.

Table 1. Cluster information

Cluster index	1	2	3	4	5	6	7	8	9	10
Number of containers	93	126	17	6	4	14	13	8	5	11

4.2   Experiment Setting

Baselines. We compare Performer with two baselines, Informer [24] and Logspare [23]. These two methods are the best Transformer-based forecasting model in time-series area so far. The general approach for data centers to perform prediction is to use one model for one cluster, that is, each cluster has its own prediction model. We follow this paradigm to train and test a prediction model for each cluster while experimenting with baseline methods.

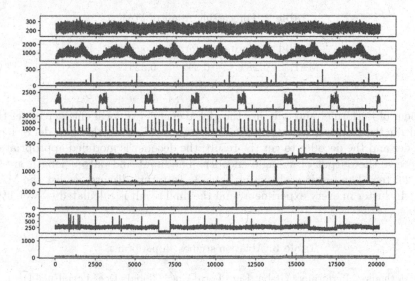

Fig. 4. Centroid of 10 container

Hyper-parameter Details. We train the model with MSE loss, optimizing with Adam optimizer. The learning rate is set as 10^{-4} and has decaying and early stopping. The batch size is 32 and the training epochs are 8. We set other baseline methods as recommended.

Metrics. We use MAE as metrics which defined as $MAE = \frac{1}{n}\sum_{i=1}^{n}|y - \hat{y}|$ where $\hat{y}$ is the predicted value, y is the ground truth and n is the prediction length. The lower MAE means better prediction accuracy.

All models are trained and tested on one NVIDIA 2070 8GB GPU. All experiments are performed 5 times and averaged.

4.3 Main Results

We compare Performer with two baseline methods on the clustered online tasks dataset. For Informer and LogTrans, we train and test 10 models for 10 cluster, as people usually do in the data center. Following the common set in long-term time-series forecasting, the input length is 96 while the output length is 24. As data centers concern more about the average performance of forecasting methods, we use the average prediction MAE of each container as the final result. From Table 2, we can see that Performer outperforms the other two baselines.

4.4 Ablation Studies

To further explore the effectiveness of each part of Performer, we experiment with several variants of it. In the first two sets of experiments, we remove the

Table 2. Forecasting result of MAE

Method	Performer	Informer	LogTrans
MAE	**0.281**	0.295	0.323

decoder of Performer and only use the global encoder or local encoder to get the result. In the next two sets of experiments, we use only the global encoder or local encoder and the decoder to get the result, the decoder is modified appropriately to receive the output of a single encoder. The input and output length are 96 and 24, too. The result is shown in Table 3. To future discuss, we list the MAE of each cluster in every experiment, and the final result is calculated by weighted average.

Table 3. Ablation studies on performer

Methods	Performer	Global Enc	Local Enc	Global+Dec	LocalEnc+Dec
Cluster 1	**0.309**	0.336	0.311	0.319	0.314
Cluster 2	**0.248**	0.248	0.261	0.250	0.254
Cluster 3	**0.240**	0.290	0.276	0.267	0.264
Cluster 4	0.106	0.182	0.106	0.114	**0.101**
Cluster 5	**0.307**	0.394	0.313	0.359	0.307
Cluster 6	**0.427**	0.429	0.433	0.431	0.435
Cluster 7	**0.283**	0.304	0.310	0.285	0.289
Cluster 8	**0.257**	0.263	0.278	0.262	0.262
Cluster 9	**0.182**	0.224	0.188	0.199	0.183
Cluster 10	0.460	0.841	0.477	0.676	**0.433**
Total	**0.281**	0.311	0.292	0.296	0.286

Firstly, we can see that Performer gains the best result in most clusters and the final result, which proves the effectiveness of Performer's design. Besides, the local encoder and decoder set is secondly good, and it gains better results than Performer in clusters 4 and 10. In the view of personalization learning, it shows that although introducing a global model can improve the overall prediction accuracy, it can be detrimental for some clients, which is common in federated personalization learning. Besides, the local encoder experiment gains better accuracy than the baseline methods mentioned above, which proves the effectiveness of our encoder's design.

5   Conclusion

In this paper, we propose Performer, a Transformer-based model for predicting future resource demands of tasks in data centers. By using an encoder-decoder architecture to combine global and local models, Performer improves the efficiency of the model using clustered datasets and improves forecasting accuracy. By using a splitting method to calculate self-attention, Performer improves its computing efficiency to be more suitable for actual deployment in data centers. Experiments show Performer is an effective method for forecasting massive clustered time-series, and it outperforms other methods.

Acknowledgement. This research is partially supported by Guangdong Natural Science Foundation of China (2018B030312002).

References

1. Tang, X., Liao, X., Zheng, J., Yang, X.: Energy efficient job scheduling with workload prediction on cloud data center. Clust. Comput. **21**(3), 1581–1593 (2018). https://doi.org/10.1007/s10586-018-2154-7
2. Cortez, E., Bonde, A., Muzio, A., Russinovich, M., Fontoura, M., Bianchini, R.: Resource central: understanding and predicting workloads for improved resource management in large cloud platforms. In: Proceedings of the 26th Symposium on Operating Systems Principles, pp. 153–167 (2017)
3. Islam, S., Keung, J., Lee, K., Liu, A.: Empirical prediction models for adaptive resource provisioning in the cloud. Futur. Gener. Comput. Syst. **28**(1), 155–162 (2012)
4. Yu, Y., Jindal, V., Yen, I.L., Bastani, F.: Integrating clustering and learning for improved workload prediction in the cloud. In: 2016 IEEE 9th International Conference on Cloud Computing (CLOUD), pp. 876–879 (2016)
5. McMahan, B., Moore, E., Ramage, D., Hampson, S., Arcas, B.A.: Communication-efficient learning of deep networks from decentralized data. In: Artificial Intelligence and Statistics, pp. 1273–1282. PMLR (2017)
6. Kulkarni, V., Kulkarni, M., Pant, A.: Survey of personalization techniques for federated learning. In: 2020 Fourth World Conference on Smart Trends in Systems, Security and Sustainability (WorldS4), pp. 794–797. IEEE (2020)
7. Box, G.E., Jenkins, G.M.: Some recent advances in forecasting and control. J. R. Stat. Soc. C Appl. Stat. **17**(2), 91–109 (1968)
8. Zivot, E., Wang, J.: Vector autoregressive models for multivariate time series. In: Modeling Financial Time Series with S-PLUS®, pp. 385–429 (2006). https://doi.org/10.1007/978-0-387-32348-0_11
9. Chen, T., Guestrin, C.: XGBoost: a scalable tree boosting system. In: Proceedings of the 22nd ACM SIGKDD International Conference on Knowledge Discovery and Data Mining, pp. 785–794 (2016)
10. Hochreiter, S., Schmidhuber, J.: Long short-term memory. Neural Comput. **9**(8), 1735–1780 (1997)
11. Box, G.E., Jenkins, G.M., Reinsel, G.C., Ljung, G.M.: Time Series Analysis: Forecasting and Control. Wiley, Hoboken (2015)
12. Rzadca, K., et al.: Autopilot: workload autoscaling at google. In: Proceedings of the Fifteenth European Conference on Computer Systems, pp. 1–16 (2020)
13. Ca Lheiros, R.N., Masoumi, E., Ranjan, R., Buyya, R.: Workload prediction using ARIMA model and its impact on cloud applications' QoS. IEEE Trans. Cloud Comput. **3**(4), 449–458 (2015)

14. Breiman, L.: Random forests. Mach. Learn. **45**, 5–32 (2001). https://doi.org/10. 1023/A:1010933404324

15. Song, B., Yu, Y., Zhou, Yu., Wang, Z., Du, S.: Host load prediction with long short-term memory in cloud computing. J. Supercomput. **74**(12), 6554–6568 (2017). https://doi.org/10.1007/s11227-017-2044-4

16. Lai, G., Chang, W.C., Yang, Y., Liu, H.: Modeling long-and short-term temporal patterns with deep neural networks. In: The 41st International ACM SIGIR Conference on Research & Development in Information Retrieval, pp. 95–104 (2018)

17. Vaswani, A., et al.: Attention is all you need. In: Advances in Neural Information Processing Systems, pp. 5998–6008 (2017)

18. Wu, N., Green, B., Ben, X., O'Banion, S.: Deep transformer models for time series forecasting: the influenza prevalence case. arXiv preprint arXiv:2001.08317 (2020)

19. Song, H., Rajan, D., Thiagarajan, J.J., Spanias, A.: Attend and diagnose: clinical time series analysis using attention models. In: Proceedings of the Thirty-Second AAAI Conference on Artificial Intelligence, vol. 32, No. 1 (2018)

20. Zerveas, G., Jayaraman, S., Patel, D., Bhamidipaty, A., Eickhoff, C.: A transformer-based framework for multivariate time series representation learning. In: Proceedings of the 27th ACM SIGKDD Conference on Knowledge Discovery & Data Mining, pp. 2114–2124 (2021)

21. Lim, B., Arık, S.Ö., Loeff, N., Pfister, T.: Temporal fusion transformers for interpretable multi-horizon time series forecasting. Int. J. Forecast. **37**(4), 1748–1764 (2021)

22. Ma, J., Shou, Z., Zareian, A., Mansour, H., Vetro, A., Chang, S.F.: CDSA: cross-dimensional self-attention for multivariate, geo-tagged time series imputation. arXiv preprint arXiv:1905.09904 (2019)

23. Li, S., et al.: Enhancing the locality and breaking the memory bottleneck of transformer on time series forecasting. Adv. Neural. Inf. Process. Syst. **32**, 5243–5253 (2019)

24. Zhou, H., et al.: Informer: beyond efficient transformer for long sequence time-series forecasting. Proc. AAAI Conf. Artif. Intell. **35**(12), 11106–11115 (2021)

25. Mansour, Y., Mohri, M., Ro, J., Suresh, A.T.: Three approaches for personalization with applications to federated learning. Comput. Sci. arXiv preprint arXiv:2002.10619 (2020)

26. Wang, K., Mathews, R., Kiddon, C., Eichner, H., Beaufays, F., Ramage, D.: Federated evaluation of on-device personalization. arXiv preprint arXiv:1910.10252 (2019)

27. Smith, V., Chiang, C.K., Sanjabi, M., Talwalkar, A.S.: Federated multi-task learning. Adv. Neural Inf. Process. Syst. **30** (2017)

28. Finn, C., Abbeel, P., Levine, S.: Model-agnostic meta-learning for fast adaptation of deep networks. In: International Conference on Machine Learning, pp. 1126–1135. PMLR (2017)

Optimizing Video QoS for eMBMS Users in the Internet of Vehicles

Lu Wang$^{(\boxtimes)}$ and Fang Fu

School of Physics and Electronic Eng, Shanxi University, Taiyuan, Shanxi, China
dearwang.lu@163.com

Abstract. Live streaming services inside the Internet-of-Vehicles (IoV) are becoming more significant in vehicle entertainment systems as a result of the fast-paced growth of the automotive industry and communications technologies. At the same time, with the rapid development of in-vehicle video, the development of broadcasting business has been driven. The 3GPP standardization group has suggested the evolved Multimedia Broadcast Multicast Service (eMBMS), which by multicasting video services to vehicle users, can significantly enhance the utilization of spectrum resources and improve signal quality. Real-time video is sent by eMBMS over synchronized spectral resources of nearby roadside units (RSUs) using a single frequency networks (SFN). However, the video transmission rate depends on the vehicle users with the lowest channel state information, to overcome these limitations, the vehicle users can be divided into groups in the SFN. It is incredibly difficult to maximize the user quality of service (QoS) of a vehicle due to dynamic nature of IoV's radio resources and channels. To overcome this obstacle, we employ a model SFN video transmission strategy that, while grouping vehicles, jointly optimizes group-oriented bit rate selection and resource allocation to decrease latency and bit rate switch in IoV. The Markov decision process (MDP), which takes into account the time-varying features of wireless channels, is used to describe the joint optimization problem. Then after, the abovementioned MDP is handled by using soft actor-critic (SAC) algorithm. The proposed approach may substantially enhance video quality while reducing latency and bit rate switch, according to extensive simulation findings based on actual data. It also works well in terms of learning efficiency and stability.

Keywords: Internet-of-vehicles · eMBMS · SFN · QoS

1 Introduction

The advancement of AI [1] and 5G technology has accelerated recently, high-definition video streaming over cellular networks is becoming more and more prevalent [2] as live video services such Douyin, Kuaishou, and Weibo gain popularity. In congested areas, resource allocation issues become more complex when numerous vehicles simultaneously request the same streaming video content, increasing peak bandwidth requirements. When popular video content is delivered using normal unicast protocols [3] in this scenario, there are few available resources and the user experience when watching videos is terrible [4].

Using the evolved multimedia broadcast multicast service (eMBMS), a 3GPP standard for 4G and 5G networks [5], provides a reliable approach to broadcast real-time video content to many cellular network users. Multicasting across the wireless spectrum is made possible by eMBMS, which groups users who are viewing the same content and broadcasting it to a group at once to maximize resource utilization. This enables effective spectrum utilization, especially when there are a large number of active users. Additionally, eMBMS improves channel conditions for vehicles by enabling roadside units (RSU) in a small area to simultaneously transmit video with the same frequency and time. Synchronized RSUs make up a RSU group, also known as a single frequency network (SFN) [6]. The signal-noise ratio (SNR) of users who are interested in this is improved by combining the signals they get from each RSU in the SFN. All RSUs and eMBMS users within an SFN have their resource allocation controlled by an multicast coordination entity (MCE).

To ensure that all vehicles receive video correctly, modulation and coding scheme (MCS) [7] are limited to vehicles with the lowest channel conditions, which affects the video viewing experience of vehicles with good channel quality. Similar limitations are imposed on synchronizing RSUs in SFN: when building vehicle groups, vehicles of one RSU with a low MCS value may adversely affect vehicles of high channel quality. It follows that vehicles should be divided into groups based on the channel conditions they are on, and each group receives an appropriate video bit rate. Vehicle grouping is therefore a vital strategy when eMBMS resource allocation is optimized.

The amount of resources allocated to each vehicle group is also a critical strategy and is limited by the resources available to eMBMS and the channel conditions of each vehicle.

In order to improve the QoS when grouping vehicles, we propose joint optimization problems on resource allocation and bitrate selection. QoE includes the video bitrate received by the vehicle, the video bitrate switching they may face, and the delay in playing the video [8].

Here is the rest of the piece. The system model is introduced in the second section, including LTE physical layer, multicast grouping model, resource scheduling and bit rate decision model, wireless transmission model, and model these problem as MDP. This article adopts the Soft Actor-Critic algorithm to solve the MDP. Section 3 presents analysis of simulation results. Section 4 provides summary and outlook.

2   Scene and Model

2.1   System Scene

In this paper, a video multicast transmission scheme based on eMBMS-SFN is proposed. As shown in Fig. 1, it consists of three RSUs and a number of vehicles driving on the road, in which the vehicles can communicate reliably with the RSUs through wireless channels, and the areas where these RSUs are located constitute a multicast broadcast single frequency network (MBSFN) area. All RSUs in SFN simultaneously multicast the same video content on the same frequency. The real-time traffic video is provided by a dynamic adaptive streaming over HTTP (DASH) source server, encoded as a set of bit rates, and multicast to vehicles on the ground via eMBMS. Vehicles interested in multicast periodically report channel conditions to the RSU, through which it is forwarded to the multicast coordination entity (MCE). Therefore, these multicast services are provided by MCE, which makes spectrum resource allocation decisions and video bit rate selection decisions for all vehicles within its coverage. Let $\mathcal{B} = \{1, \cdots, B\}$ denote the set of roadside units RSU, and $\mathcal{M} = \{1, \cdots, M\}$ denote the set of eMBMS vehicles, and $\mathcal{N} = \{1, \cdots, N\}$ denote the set of RB numbers.

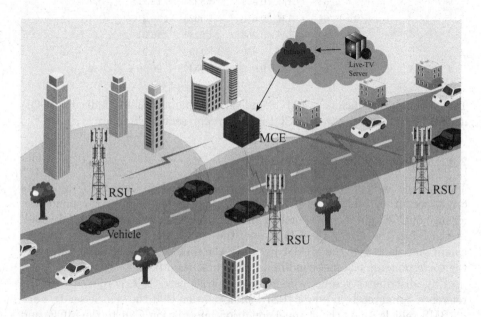

Fig. 1. System scene

The number of MCS and RBs that transmit the bit rate, are what we refer to as a complete network configuration.

2.2 LTE Physical Layer

OFDMA technology is the foundation of the LTE downlink's physical layer. The fundamental unit of resources in the LTE system is a physical resource block (RB), which is a time slot (0.5 ms) with a bandwidth of 180 KHz, 12 subcarriers (180 kHz) and 7 symbols [9]. All subcarriers within an RB use the same modulation and coding scheme (MCS). Therefore, the number of bits that an RB may carry will be determined by the MCS of an RB [10], i.e.:

$$c(MCS) = 12(subcarries) \times 7(symbols) \times efficiency, \tag{1}$$

Table 1. MCS Table

CQI	modulation	code rate	efficiency	SNR
1	$QPSK$	0.077	0.1523	−6.06
2	$QPSK$	0.117	0.2344	−4, 54
3	$QPSK$	0.188	0.3770	−2.78
4	$QPSK$	0.301	0.6016	−0.97
5	$QPSK$	0.438	0.8770	0.92
6	$QPSK$	0.588	1.1758	2.79
7	$16 - QAM$	0.369	1.4766	4.72
8	$16 - QAM$	0.479	1.9141	6.64
9	$16 - QAM$	0.602	2.4603	8.47
10	$64 - QAM$	0.455	2.7305	10.46
11	$64 - QAM$	0.554	3.3223	12.27
12	$64 - QAM$	0.650	3.9023	14.21
13	$64 - QAM$	0.754	4.5234	16.01
14	$64 - QAM$	0.853	5.1152	17.95
15	$64 - QAM$	0.926	5.5547	20.80

Table 1 shows the efficiency of MCS and various channel quality indicator (CQI) metrics. In this article, we use CQI index to represent MCS index.

2.3   Multicast Grouping Model

This scheme assigns vehicles to different groups based on their SNR values. As can be seen from Table 1, the SNR and CQI value are one-to-one correspondence, a range of CQI values correspond to an MCS, there are a total of 15 CQI values in the table, the larger the CQI value, the better the corresponding channel conditions, the better the MCS, conversely, the smaller the CQI value, the worse the corresponding channel conditions, the worse the MCS.

RSU distributes resources to related unicast users in accordance with the reference signal received power (RSRP) information that the vehicle sends. Each eMBMS vehicle sends the channel condition information CQI to the MCE, and the MCE divides the eMBMS vehicles into K groups according to the similarity of the CQIs.

The vehicles in the eMBMS group all receive video through the same group of RBs, let $\mathcal{K} = \{1, \cdots, K\}$ denote the set of group. To guarantee that all vehicles in the k group can receive the signal successfully, the MCS of each group depends on the vehicle with the lowest channel state information. According to the mapping relationship table between SNR and CQI, the corresponding CQI can be found by calculating the SNR, then during multicast transmission at time t, we can express the SNR [11] of the eMBMS vehicle m as:

$$\gamma_m(t) = \frac{1}{\sigma^2} \left| \sum_{b \in B} h_{m,b}(t) \right|^2 \tag{2}$$

where $h_{m,b}(t)$ is the complex channel coefficient, which can be expressed as:$h_{m,b}(t) = \gamma_b \cdot \tilde{h}_{m,b}(t)$, where γ_b represents the macroscopic path loss and shadow fading, $\tilde{h}_{m,b}(t)$ indicates microscopic fading, σ^2 is additive white Gaussian noise.

2.4   Resource Scheduling and Bit Rate Decision Models

Unicast vehicles take up X of the N available RBs per RSU b, while eMBMS vehicle m uses the remaining RBs. The number of resources that RSU b can be assigned to eMBMS vehicle m is represented by ξ. The size of the ξ is equal to the minimum value between the remaining RBs number $(N - X)$ and the total RB number of $\beta(\beta N)$ [12], which can be expressed as:

$$\xi = min(\beta N, N - X) \tag{3}$$

where β is usually set to 60% [12].

The DASH source server encodes the video into different versions of I, each version of $i \in I$, corresponding to a video bitrate v_i. The number of RBs required per version i depends on the MCS. The MCE can select to transmit one or more various bit rates, each of which corresponds to a different MCS modulation, depending on the channel conditions of the vehicle. The appropriate bit rate is selected according to the selected MCS modulation. The number of RBs required for this bit rate depends on the MCS.

We use $J(t)$ to define the group-oriented video version selection strategy, which is an $\mathcal{K} \times \mathcal{I}$ matrix with binary variables $z_{k,i}$, in addition, since each video version can only be used by at most one group selection in each time period, the variable $z_{k,i}$ satisfies:

$$\sum_{i=1}^{I} z_{k,i}(t) = 1, z_{k,i}(t) \in \{0, 1\}, \forall k \in \mathcal{K}. \tag{4}$$

where $z_{k,i}(t) = 1$ means that the video version i is chosen to be used by the vehicles of the k group, otherwise $z_{k,i}(t) = 0$.

In the t time slot, the video bitrate selected for group k is:

$$v_k(t) = \sum_{i=1}^{I} z_{k,i}(t) v_i \tag{5}$$

2.5   Wireless Transmission Model

The video should be multicast in accordance with the lowest channel state informations of the vehicles in the multicast group to guarantee that every vehicle in the group can correctly decode the received video. Consequently, the SNR for the k group is:

$$\gamma_k(t) = \min_{m \in M} \gamma_m(t) \tag{6}$$

We use Shannon capacity to calculate the maximum download rate of group k on RB n, i.e.:

$$R_{b2k}(t) = \omega log_2(1 + \gamma_k(t)) \tag{7}$$

where ω is the bandwidth of a single RB.

We use $\boldsymbol{H(t)}$ to define the group-oriented resource allocation policy using h, which is an $\mathcal{K} \times \mathcal{N}$ matrix with binary variables $\mu_{k,n}$, in addition, since each RB can only be occupied by at most one group in each time period, the variable $\mu_{k,n}$ satisfies:

$$\sum_{k=1}^{K} \mu_{k,n}(t) = 1, \mu_{k,n}(t) \in \{0,1\}, \forall n \in \mathcal{N}. \tag{8}$$

where $\mu_{k,n}(t) = 1$ denotes that RB n is assigned to group k at slot t, and $\mu_{k,n}(t) = 0$ otherwise.

All vehicles in group k share the same channel, so they all experience the same download rate in each time period. The instantaneous speed of the vehicle in group k is given by:

$$R_k(t) = \sum_{n=1}^{N} \mu_{k,n}(t) R_{b2k}(t). \tag{9}$$

The size of the video to be sent by RSU b can be calculated as:

$$\Xi(v_i) = \tau_s v_k(t) \tag{10}$$

where τ_s is the playback time of the video.

The transmission delay for RSU b to send video to group k vehicles is:

$$T_k(t) = \frac{\Xi(v_k)}{R_k(t)} \tag{11}$$

In addition, in order to ensure that the vehicle can successfully receive the video, the selected video bitrate cannot exceed the maximum download rate of each group, i.e.:

$$v_k \leq R_k \tag{12}$$

2.6   Problem Formulation

Video Quality of eMBMS Vehicle: In SFN, we define it as the bit rate utility of the video v of the eMBMS vehicle M, according to [13], which can be expressed as:

$$Q(M_v) = \sum_{v \in v_i} \sum_{m \in M_v} ln(1 + \frac{v_i(t)}{v_i(t-1)}) U_{v,i}^m \tag{13}$$

where $v_i(t)$ and $v_i(t-1)$ are the video bit rates for time slot t and time slot $(t-1)$, respectively. $U_{v,i}^m$ is a binary variable that indicates whether the vehicle m in the SFN interested in the video v should receive the bit rate v_i.

Bitrate Switching of eMBMS Vehicle: Frequently switching video versions will seriously affect the viewing experience of vehicle users, and bitrate switching is taken as a penalty item in the quality of user experience to avoid bitrate switching as much as possible, i.e.:

$$V(t) = \sum_{v \in v_i} \sum_{m \in M_v} |v_i(t) - v_i(t-1)| \cdot U_{v,i}^m \tag{14}$$

Time Delays of eMBMS Vehicle: The video transmission time consumption from RSU B to vehicles involves of two periods: video transmission time T_{b2k} and decoding time T_d from RSU b to K group vehicles. In order to ensure smooth video playback, $T_{b2k}(t) + T_d \leq \tau_s$ needs to be satisfied, otherwise there will be delays:

$$D(t) = \sum_{k \in K} \sum_{m \in M_v} (T_k(t) + T_d - \tau_s) \cdot U_{v,i}^m \tag{15}$$

Utility Function: Considering video quality, bit rate switching and delay, the utility function is defined as:

$$\max_{J(t), H(t)} \varphi Q(t) - \beta V(t) - \omega D(t)$$
$$s.t.(4), (8), (12) \tag{16}$$

where $\varphi(\$)$ is the reward price of video quality, $\beta(\$/bits/second)$ is the penalty price caused by bit rate switching, and $\omega(\$/second)$ is the penalty price caused by delay. So the utility function is finally expressed as the size of the payoff($\$$) [14].

2.7   Problem Modeling Based on Markov Decision Process

Due to their Markov properties [15] such as channel conditions, vehicle driving positions and RBs, In this subsection, the joint optimization problem is modeled as MDP, where the MDP consists of $< S, A, r >$: S represents the state space, which involves the channel gain $h_{mb}(t)$ between the RSU b and the vehicle m, the current video version bit rate and available spectrum resources; A represents the action space, which contains the resource allocation strategy and the video version selection strategy, and r represents the reward size of the environment feedback to the agent after performing an action.

In reinforcement learning, the optimal strategy $\mu(a_t|s_t)$ is found by continuously training the agent, thereby maximizing the cumulative reward. Among them, policy $\mu(a_t|s_t)$ represents the probability of action a made when the state is s. After taking an action a, the environment will immediately return to reward r, which is expressed as:

$$r = \{\varphi Q(t) - \beta V(t) - \omega D(t)\} \tag{17}$$

Next, the advanced DRL algorithm will be used to solve the above problems.

3   Problem Solving with Soft Actor-Critic Algorithm

This study adopts the advanced SAC algorithm to solve the above MDP problem. The traditional RL algorithm is to observe the action taken by the environment according to the agent, output the next state and the reward brought by the current action, and constantly update the strategy to maximize the expected reward value. The ultimate goal of the algorithm is to maximize the expected value reward and entropy. Compared with the RL algorithms, the adopted SAC DRL algorithm has the following advantages. The first is an off-policy algorithm that maximizes the entropy on the basis of maximizing the reward value to formulate random strategies, the second is to have stronger reward exploration and stability. Therefore, it can be used in a more complicated multicast environment.

3.1   Soft Value Function

Reinforcement learning generally solve the state value function through iterative Belman equations, and finally convergence the state action value function through continuous iteration, so as to obtain the optimal strategy. The difference between the existing RL algorithm [16] is that the entropy term $H(\mu(a_t|s_t))$ [17] is added to the SAC algorithm. The SAC algorithm has maximized the entropy of the strategy while obtaining the maximum reward, which can enhance the exploratory nature of the agent. The optimal strategy$\mu(a_t|s_t)$ is expressed as:

$$\mu^* = arg\,\underset{\mu}{max}\,\mathbb{E}_{(s_t,a_t)\sim\rho_\mu}\left\{\sum_{t=0}^{\infty}\lambda^t\left[r(s_t,a_t) + \beta H\left(\mu(a_t\mid s_t)\right)\right]\mid\mu\right\} \tag{18}$$

where $H(\mu(a_t|s_t)) = -log\mu(a_t|s_t)$ is entropy term, λ is the discount factor, which is used to reflect the impact of the current reward on future rewards, and β is the temperature parameter, which controls the degree of randomness of the strategy. When given an initial state s_t and an initial action a_t, the entropy expectation is expressed as:

$$Q^\mu(s_t,a_t) = \mathbb{E}_{(s_t,a_t)\sim\rho_\mu}\left\{\sum_{t=0}^{\infty}\lambda^t\left[r(s_t,a_t) + \beta H\left(\mu(a_t\mid s_t)\right)\right]\mid s_0 = s_t, a_0 = a_t, \mu\right\} \tag{19}$$

Theorem 1. *The state value function V can be expressed as a Q value function as:*

$$V^\mu(s_t) = \beta log \int_A exp(\frac{1}{\beta}Q^\mu(s_t,a_t))da \tag{20}$$

The corresponding optimal strategy is:

$$\mu'(\cdot \mid s_t) = \exp\left(\frac{1}{\beta}Q^\mu(s_t,a_t) - V^\mu(s_t)\right) \tag{21}$$

where the Q value and the state value correspond to the soft Q value and the soft state value in the SAC algorithm.

3.2 Critic Section

In the actor-critic algorithm framework, critic uses the previous value function to analyze the actor's performance and lead the actor's next stage of action.

The random strategy makes the SAC algorithm have stronger exploration capabilities, similar to other RL algorithms, the Q-value function Q in the critical part parameterizes the Q-value by using a set of deep neural networks (DNNs) [18] with weights $\eta = \{\eta_1, \eta_2, \cdots, \eta_N\}$, i.e. $Q(s_t, a_t) \approx Q_\eta(s_t, a_t)$.

The soft Q-value is trained by calculating the gap between the outputs of the critic network for action a as the loss function:

$$J_Q(\eta) = \mathbb{E}_{(s_t,a_t)\sim\rho_\mu}\left[\frac{1}{2}(Q_\eta(s_t,a_t) - \hat{Q}(s_t,a_t))^2\right] \tag{22}$$

where $\hat{Q}(s_t, a_t)$ satisfies $\hat{Q}(s_t, a_t) = r(s_t, a_t) + \lambda \hat{V}_{\hat{\eta}}(s_{t+1})$, and $\hat{V}_{\hat{\eta}}(s_{t+1})$ is the target state value, and a set of samples (s, a, r, s_{t+1}) are obtained from the experience pool. The parameter η is updated by gradient descent, which is expressed as:

$$\eta(t+1) \leftarrow \eta(t) - \alpha_{c,t}\nabla_\eta J_Q(\eta) \tag{23}$$

where $\alpha_{c,t}$ is the critic learning rate, and the gradient of $J_Q(\eta)$ can be expressed as:

$$\nabla_Q J_Q(\eta) = \nabla_\eta Q_\eta(s_t,a_t)\left[Q_\eta(s_t,a_t) - r(s_t,a_t) - \lambda\hat{V}_{\hat{\eta}}(s_{t+1})\right] \tag{24}$$

Similarly, the state value function in the SAC algorithm is parameterized by another set of DNNs with weight of $\iota = \{\iota_1, \iota_2, \cdots, \iota_N\}$, i.e. $V(s) = V_\iota(s)$. The soft V-value is trained by calculating the gap between the outputs of the critic network for action a as the loss function:

$$J_V(\iota) = \mathbb{E}_{(s_t,a_t)\sim\rho_\mu}\left[\frac{1}{2}(V_\iota(s_t) - E[Q_\eta(s_t,a_t) + \alpha H(\mu_\varsigma(a_t|s_t))])^2\right] \tag{25}$$

The corresponding $J_V(\iota)$ gradient is:

$$\nabla_V J_V(\iota) = \nabla_\iota V_\iota(s_t)[V_\iota(s_t) - [Q_\eta(s_t,a_t) + \alpha H(\mu_\varsigma(a_t \mid s_t))]] \tag{26}$$

The update method of parameter ι is:

$$\iota(t+1) \leftarrow \iota(t) - \alpha_{c,t}\nabla_{\iota}J_V(\iota) \tag{27}$$

The parameter of the corresponding target state value function $\hat{V}_{\bar{\iota}}(s_{t+1})$ is $\iota = \{\iota_1, \iota_2, \cdots, \iota_N\}$. The smooth update of parameter $\bar{\iota}$ is:

$$\bar{\iota}(t+1) \leftarrow \tau\iota(t) + (1-\tau)\bar{\iota}(t) \tag{28}$$

where τ is a smoothing factor that satisfies $0 < \tau < 1$.

3.3   Actor Section

In the framework of the actor-critic algorithm, actor use policy functions to perform actions as outputs to the environment.

The parameter of the actor network is ζ, input state s_t and output action a_t. Different from other algorithms, the actor network of the SAC algorithm outputs a Gaussian distribution, which contains mean and standard deviation, and uses the Kullback-Leibler (KL) divergence to minimize the following expected values to obtain the optimal strategy:

$$J_{\mu}(\zeta) = \mathbb{E}_{(s_t,a_t)\sim\rho_{\mu}}\left(D_{KL}\mu_{\zeta}(\cdot \mid s_t)\|\mu'(\cdot \mid s_t)\right) \tag{29}$$

This paper reparameterizes the policy, i.e. $a = f_{\zeta}(\epsilon_t, s_t)$, where ϵ is the sampled motion noise value. Then $J_{\mu}(\zeta)$ is expressed as:

$$J_{\mu}(\zeta) = \mathbb{E}_{(s_t,a_t)\sim\rho_{\mu}}\left[\beta H(\mu_{\zeta}\left(f_{\zeta}(\epsilon_t, s_t)\right) \mid s_t) - Q^{\mu}\left(s_t, f_{\zeta}(\epsilon_t, s_t)\right) + V^{\mu}(s_t)\right] \tag{30}$$

The update process of the actor target network parameters is to minimize the KL divergence. The smaller the KL, the smaller the difference between the reward corresponding to the output action of each time, and the convergence result can be obtained. In the end, the gradient ζ is expressed as:

$$\nabla_{\zeta}J_{\mu}(\zeta) = \nabla_{\zeta}\beta H(\mu_{\zeta}(a_t \mid s_t)) - \nabla_a\beta H(\mu_{\zeta}(a_t \mid s_t)) - \nabla_aQ_{\eta}(s_t, a_t)\nabla_{\zeta}f_{\zeta}(\epsilon_t, s_t) \tag{31}$$

The strategy parameter ζ is updated by using a gradient descent method, which is expressed as:

$$\zeta(t+1) \leftarrow \zeta(t) - \alpha_{a,t}\nabla_{\mu}J_{\mu}(\zeta) \tag{32}$$

where $\alpha_{a,t}$ represents the actor learning rate.

Theorem 2. *The SAC algorithm achieves convergence by alternately performing critic and actor. For any $\forall\mu, \mu^* \in \prod(\mu^* \neq \mu)$ and $\forall(s_t, a_t) \in S\times A$, satisfying $Q^{\mu^*}(s_t, a_t) > Q^{\mu}(s_t, a_t), \forall\mu \in \prod$, converges to policy $\mu^*(\cdot|s_t)$.*

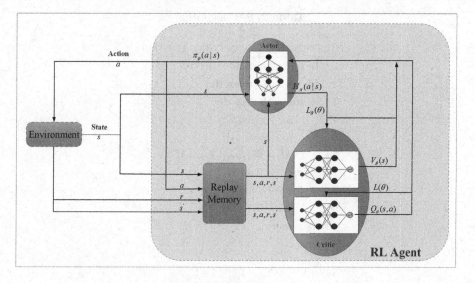

Fig. 2. SAC algorithm

3.4   Soft Actor-Critic DRL Algorithm

Figure 2 depicts the architecture of SAC based algorithm. As can be seen from the figure, the SAC algorithm contains an actor network and two critical networks, and adopts experience replay mechanism.

SAC algorithm network model is composed of multiple networks during iteration: the strategy network make corresponding strategy according to that environment state and through its network layer distribution state of the output behavior to compute the mean and standard deviation of action at the moment; soft Q network obtain an expected Q value according to that current action and state, value network evaluates the strategy based on the state; target value network is a copy network of the value network. Its network parameters and value are the same, but the update speed is different from the value network. In the implementation of the SAC algorithm model, this algorithm can be trained and learned from past experience, and the past experience sample data is stored in the replay memory, take a small sample for use in that algorithm. Then update the parameters based on the samples extracted from the experience pool, until this round is terminated; then the algorithm selects the data of the entire turn to update the entire round to make strategic updates, finally, the training is complete. The algorithm is detailed in Table 2.

4   Analysis of Simulation Results

The simulation is implemented with $Tensor flow$ 1.4.0, and the hardware environment is based on an x64 processor laptop with Intel(R) Core(TM) i7-6500U CPU and 224 G memory. The simulation software uses a python simulator, and

Table 2. SAC algorithm

Algorithm: Soft actor-critic algorithm
1   Initialize parameter $\hat{\iota}, \iota, \eta, \zeta$;
2   **for** *each iteration* **do**;
3   **for** each step **do**
4           $a_t \sim \mu(\cdot \vert s_t, \zeta)$;
5           $s'_t \sim p(s_t, a_t)$;
6           $M \leftarrow (s_t, a_t, r, s'_t) \cup M$
7   end
8   **for** *each update step* **do**;
9   (23),(27),(28),(32)
10   end
11 end

the environment is constructed to deploy 3 RSUs and 5 vehicles on an urban road. This article uses DASH video, and the specific parameters are set in Table 3.

Table 3. Simulation parameter

Parameter	Data
Number of RSUs (or base stations)	3
Number of vehicles	5
RB resource block	$180\,kKz$
Path loss index	1.42
Video segment duration	$1s$

Figure 3 shows the average rewards obtained under different actor learning rates when $\alpha_{c,t} = 0.02$. As can be seen from the figure, when $\alpha_{a,t} = 0.0005$, compared with $\alpha_{a,t} = 0.0001$, the convergence speed is far exceeded. In the figure, the convergence has been reached around 200 rounds, and the reward value is higher than when $\alpha_{a,t} = 0.0001$ reward. When $\alpha_{a,t} = 0.019$, although the convergence speed of the two is basically the same, when $\alpha_{a,t} = 0.019$, the fluctuation after convergence is larger, and the overall convergence reward is much lower than that when $\alpha_{a,t} = 0.0005$. Therefore, the learning effect is optimal when the actor learning rate is 0.0005, and the learning rate is set to $\alpha_{a,t} = 0.0005$ in the following simulations.

Figure 4 shows the average rewards obtained under different critic learning rates when $\alpha_{a,t} = 0.0005$. It can be seen from the figure that when $\alpha_{c,t} = 0.02$, the convergence speed is faster than when $\alpha_{c,t} = 0.0007$, and the convergence has been reached in about 100 rounds in the figure. When $\alpha_{c,t} = 0.07$, although the

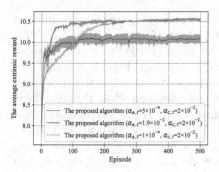 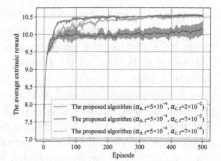

Fig. 3. Different actor learning rates   **Fig. 4.** Different critic learning rates

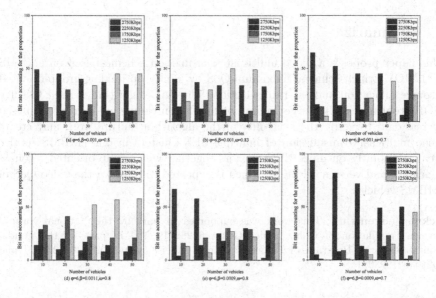

Fig. 5. Different number of vehicles

convergence rate is basically the same as that when $\alpha_{c,t} = 0.02$, the convergence value when $\alpha_{c,t} = 0.07$ is lower than that when $\alpha_{c,t} = 0.02$. Therefore, the learning effect is optimal when the critic learning rate is 0.02, and the learning rate is set to $\alpha_{a,t} = 0.0005$ and $\alpha_{c,t} = 0.02$ in the following simulations.

Figure 5 shows the proportion of the DASH video bit rate corresponding to different numbers of vehicles with the continuous increase of the number of vehicles under the condition that the bandwidth resources remain unchanged. As can be seen from the figure, the overall trend of the six figures is that as the number of vehicles increases, the proportion of high video bit rates (2750 kbps and 2250 kbps) will gradually decrease, while the proportion of low video bit rates (1750 kbps and 1250 kbps) will gradually increase. At the same time, the figure shows the impact of bit rate switching price β and delay price ω on different video bit rate ratios. Figure 3(a) shows that when $\varphi = 6$, $\beta = 0.01$ and $\omega = 0.8$, it can be seen that the proportion of low bit rates (1750 kbps and 1250 kbps) increases significantly as the number of vehicles increases. Figures 3(b) and 3(c) show that when $\varphi = 6$ and $\beta = 0.01$, it can be seen that when the time delay price ω increases, the overall video bit rate decreases, and when the time delay price ω decreases, the overall video bit rate increases. Figures 3(d) and 3(e) show that when $\varphi = 6$, $\omega = 0.8$, it can be seen that when the bit rate switching price β increases, the overall video bit rate decreases, and when the bit rate switching price β decreases, the overall video bit rate increases. Figure 3(f) shows that when $\varphi = 6$, it can be seen that when the price of bit rate switching β and the price of delay ω are reduced at the same slot, the overall video bit rate is greatly increased.

5   Summarize

This paper proposes a video multicast transmission scheme based on eMBMS in the IOV, which achieves maximum QoS by jointly optimizing group-oriented resource allocation and bit rate selection. The SAC algorithm is used to solve MDP problem. Finally, large amount of simulation results are carried out based on real data to verify the feasibility of the algorithm. Future work may create video multicast transmission of multiple SFN clusters in one eMBMS service area, and jointly optimize SFN clusters together with group-oriented resource allocation and version selection, which is expected to improve the video QoS of eMBMS vehicles.

Acknowledgements. This work was supported in part by the National Natural Science Foundation of China under Grant 62261019 and in part by the Fundamental Research Program of Shanxi Province under Grants 202103021224024 and 202103021223021.

References

1. Wu, Y., Guo, H., Chakraborty, C., et al.: Edge computing driven low-light image dynamic enhancement for object detection. IEEE Trans. Netw. Sci. Eng., 1 (2022)
2. Afolabi, R.O., Dadlani, A., Kim, K.: Multicast scheduling and resource allocation algorithms for OFDMA-based systems: a survey. IEEE Commun. Surv. Tutorials **15**(1), 240–254 (2013)
3. Boban, M., Tonguz, O.K., Barros, J.: Unicast communication in vehicular ad hoc networks: a reality check. IEEE Commun. Lett. **13**(12), 995–997 (2009)
4. Fu, F., Kang, Y., Zhang, Z., et al.: Soft actor-critic DRL for live transcoding and streaming in vehicular fog-computing-enabled IoV. IEEE Internet Things J. **8**(3), 1308–1321 (2021)
5. Shi, G., Wu, Y., Liu, J., et al.: Incremental few-shot semantic segmentation via embedding adaptive-update and hyper-class representation. In: Proceedings of the 30th ACM International Conference on Multimedia 2022, pp. 5547–5556. ACM (2022). https://doi.org/10.1145/3503161.3548218

6. Araniti, G., Rinaldi, F., Scopelliti, P., et al.: A dynamic MBSFN area formation algorithm for multicast service delivery in 5G NR networks. IEEE Trans. Wireless Commun. **19**(2), 808–821 (2020)
7. Choi, Y.I., Kang, C.G.: Scalable video coding-based MIMO broadcasting system with optimal power control. IEEE Trans. Broadcast. **63**(2), 350–360 (2017)
8. Zhang, Z., Wang, R., Yu, F.R., et al.: QoS aware transcoding for live streaming in edge-clouds aided HetNets: an enhanced actor-critic learning approach. IEEE Trans. Veh. Technol. **68**(11), 11295–11308 (2019)
9. Boni, A., Launay, E., Mienville, T., et al.: Fifth IEE international conference on 3G mobile communication technologies. 1nd edn. IEE, London (2004)
10. Park, J., Tarkhan, A., Hwang, J.N., et al.: 2017 IEEE international conference on communications (ICC). 1nd edn. IEEE, Piscataway (2017)
11. Du, J., Cheng, W., Lu, G., et al.: Resource pricing and allocation in MEC enabled blockchain systems: an A3C deep reinforcement learning approach. IEEE Trans. Netw. Sci. Eng. **9**, 33–44 (2022)
12. EBU: delivery of broadcast content over LTE networks. 1nd edn. European Broadcasting Union, Geneva (2014)
13. Reichl, P., Tuffin, B., Schatz, R.: Telecommunication Systems, 1st edn. Springer Nature, Germany (2013)
14. Zhang, Z., Yu, F.R., Fu, F., et al.: Joint offloading and resource allocation in mobile edge computing systems: an actor-critic approach. In: 18th IEEE Global Communications Conference, Abu Dhabi (2018)
15. Roy, A., Borkar, V., Karandikar, A., et al.: Online reinforcement learning of optimal threshold policies for Markov decision processes. IEEE Trans. Autom. Control **67**(7), 3722–3729 (2022)
16. Zhang, Z., Zhang, Q., Miao, J., et al.: Energy-efficient secure video streaming in UAV-enabled wireless networks: a safe-DQN approach. IEEE Trans. Green Commun. Netw. **5**(4), 1982–1995 (2021)
17. Haarnoja, T., Tang, H., Abbeel, P., et al.: Reinforcement learning with deep energy-based policies. In: ICML, vol. 70, no. 8, pp. 1352–1361 (2017)
18. Sutton, R.S., Barto, A.G.: Reinforcement Learning: An Introduction, 1st edn. MIT Press, Cambridge (2017)

Huffman Tree Based Multi-resolution Temporal Convolution Network for Electricity Time Series Prediction

Chao Tu[1], Ming Chen[1], Liwen Zhang[1], Long Zhao[2], Yong Ma[3(✉)], Ziyang Yue[3], and Di Wu[3]

[1] State Grid Jiuquan Power Supply Company, Jiuquan, China
{tuchao,chenming,zhanglw1}@gs.sgcc.com.cn
[2] Electric Power Research Institute of State Grid Gansu Electric Power Company, Lanzhou, China
zhaolong@gs.sgcc.com.cn
[3] Huazhong University of Science and Technology, Wuhan, China
{yongma2,ziyangyue,woodybryant}@hust.edu.cn

Abstract. Electricity time series prediction is a fundamental part in electricity system scheduling that maintains the balance between electrical supply and demand. However, most existing methods cannot capture the complicated structure of electricity time series, and make personalized suggestions on electricity purchasing scheme. The main challenge lies in the periodicity and instability of electricity time series. To capture the global and local features simultaneously, we propose a **Huffman Tree based Multi-resolution Temporal Convolution Network** (HM-TCN). HM-TCN can extract and integrate multi-resolution features and therefore achieve high prediction accuracy. We also implement a cost optimization framework based on HM-TCN with visualization interface. Extensive experiments with real-world data offer evidence that the proposed method outperform the baselines by reducing RMSE by at least 89%.

Keywords: Deep learning · Time series prediction · Electricity load prediction

1 Introduction

Electricity time series prediction is a fundamental part in electricity system scheduling and the basics of electricity economy. Due to the fact that electric energy cannot be stored massively and must be consumed after being generated, it is necessary to predict the electricity load accurately in order to assure the balance between supply and demand. In addition, under the background of industrial electricity marketization, users with different demand of electricity need personalized purchasing schemes. An unreasonable purchasing scheme may result in high electricity cost. For example, the difference between planned electricity consumption and actual electricity consumption gives rise to additional fee of deviation electricity assessment.

C. Yu et al. (Eds.): GPC 2022, LNCS 13744, pp. 230–245, 2023.
https://doi.org/10.1007/978-3-031-26118-3_18

According to State Grid Corporation of China (SGCC), high voltage power consumers consist of small industrial users, large industrial users, and non-industrial users. For small industrial users, electricity charge depends on the amount of electricity consumed and power rate, which is called single-part charge. For large industrial users, electricity charge depends on the amount of electricity consumed, maximum power, and power rate, which is called two-part charge. While electricity charge of non-industrial users depends only on amount of electricity consumed. Details of electricity charge and the division of users are covered in Sect. 2.

Classical methods of electricity time series prediction includes statistics based methods and machine learning based methods. Statistics based methods include Autoregressive Integrated Moving Average model (ARIMA) [8], Kalman Filter [1] and Auto Regressive Conditional Heteroscedasticity (ARCH) [13]. Machine learning based methods include neural network [20], Auto Regressive Conditional Heteroscedasticity (SVM) [14] and Random Forest (RF) [6]. While during the application of these methods, we noticed several problems:

1. Electricity time series is characterized by periodicity, instability and randomness. Existing methods cannot capture the complicated structure of time series data, which leads to poor prediction.
2. Statistic based methods are simple and easy to train, but cannot fit the electricity time series well, especially for nonlinear series. Machine learning based methods face difficulty in tuning parameters, and cannot mine the temporal correlation.
3. Existing methods do not consider the policy of electricity purchasing scheme. They cannot give personalized suggestion based on the electricity consumption characteristics of users.

To solve the problems, we propose a **H**uffman Tree based **M**ulti-resolution **T**emporal **C**onvolution **N**etwork (HM-TCN). It first performs multi-layer convolution on the time series to extract multi-resolution temporal features, which captures the characteristics of time series at different levels. Then HM-TCN fuses the multi-resolution information by Huffman tree to provide richer feature information for time prediction. Compared with classical methods, HM-TCN has stronger feature extraction ability due to its non-linear functionality and deep learning capability, which can mine the features of time series information from both short and long time ranges. The convolution structure can process data in parallel, which shortens the time of model training and model inference and is able to deal with massive datasets. In addition, we take the policy of electricity purchasing scheme into consideration and implement a cost optimization framework with visualization interface based on HM-TCN. It displays various information of electricity consumption, and gives personalized suggestion to users whose purchasing scheme is unreasonable.

The major contributions are summarized as follows:

– We study the electricity time series prediction problem under the background of industrial electricity marketization with deep learning techniques.

- We propose a Huffman Tree based Multi-resolution Temporal Convolution Network, named HM-TCN. It can extract and integrate multi-resolution features of time series. Besides, we implement a cost optimization framework. The visualization interface presents various information of electricity consumption, results of prediction, and personalized suggestion of electricity purchasing scheme.
- We report on extensive experiments on real-world electricity time series datasets. Experimental results prove that the proposed methods can predict electricity time series with high accuracy.

2   Related Work

Time series data prediction is a challenging task. So far, there have been a large number of related studies on time series data prediction. AutoRegressive Integrated Moving Average (ARIMA) [16] model the time series data by considering the moving average component, where the non-stationary time series data could be transformed into linear stationary by differentiation. However, this kind of method can only deal with linear time relationship, and can not handle nonlinear time dependencies. Besides, the parameters need to be selected manually, lacking the mechanism of automatic learning. Other statistic models like Bayesian [4], Markov [26], Holt [5] and GAF [22]. MultiLayer Perceptron (MLP) [11] can learn nonlinear and hierarchical discriminative features from multidimensional time series data, and activation functions (Sigmoid, Relu, Tanh) are employed to provide the nonlinearity. In order to capture the nonlinear time dependence between time series data, many methods based on Recurrent Neural Network (RNN) [9,10,23] have been proposed. Due to the phenomenon of gradient disappearance and explosion in traditional RNN, two variants of RNN: Long Short-Term Memory (LSTM) [10] and Gate Recurrent Unit (GRU) [7] are used to replace RNN, to extract long-term dependencies. Rahman et al. [17] implemented and optimized an RNN model, applying it to solve medium to long-term electricity load prediction at one hour resolution, and to impute electricity consumption datasets containing segments of missing values. Convolutional Neural Network (CNN) [18,21,27] can be used to extract local spatial dependencies, and 1D CNN is commonly used to extract temporal dependencies. Shaojie Bai et al. [2] proposed a model based on causal dilation convolution to replace RNN to better extract temporal dependencies. Bendaoud et al. [3] build a cGAN architecture based on the GAN model (Generative Adversarial Networks) to accurately and stably capture the periodic changes of the electrical load. Besides, there are some other deep learning structures proposed. Yirui Wu et al. [19,24,25] propose some useful deep learning methods for object detection, image encryption and segmentation.

3   Preliminary

We proceed to present the problem setting. Then we elaborate details of electricity cost computation and the division of users.

3.1 Problem Setting

Definition 1 (Time Series). *A univariate time series sequence is denoted as* $\mathcal{X} = \{X_1, X_2, ..., X_T\}$, *where* $X_t \in R, t = 1, ..., T$ *is an observation at time t and T is the length of the time series.*

Definition 2 (Electricity Time Series Prediction). *Given a univariate time series sequence* $\mathcal{X} = \{X_1, X_2, ..., X_T\}$, *where* X_t *represents the electricity load of time t. The Electricity Time Series Prediction aims to predict the electricity time series of time* $T + 1, ..., T + \Delta t$, *i.e.* $\hat{\mathcal{X}} = \{\hat{X}_{T+1}, ..., \hat{X}_{T+\Delta t}\}$, *such that* $\hat{\mathcal{X}}$ *is as similar with the real electricity time series* $\mathcal{X} = \{X_{T+1}, ..., X_{T+\Delta t}\}$ *as possible.*

Electricity Time Series Prediction can be divided into average load prediction and maximum load prediction. For average load prediction, X_t represents the electricity consumed per day. For maximum load prediction, X_t represents the power per 15 min.

3.2 Details of Electricity Cost Computation

Electricity Charge. Electricity charge equals to electricity consumed times unit price that differs in peak, plain and valley time. Peak unit price is 1.5 times of plain price, and valley unit price is 0.5 times of plain price. For marketized users, electricity charge, additional fee of deviation electricity assessment are required. The more the deviation of planned and actual electricity consumed, the more the user is charged. Therefore, accurate average load prediction helps cut down electricity charge.

Basic Electricity Charge. Large industrial users are those whose total capacity of transformers is larger than 315kVA. For these users, basic electricity charge should be taken into consideration. Basic electricity charge falls into two categories: capacity based charge and demand based charge. Demand based charge is further divided into actual maximum demand charge and contract maximum demand charge. For capacity based charge, basic electricity charge is computed as follows:

$$cost = C_{max} \times \frac{N_{on}}{N_{tot}} \times 19 \tag{1}$$

where C_{max} is the total capacity of transformers, N_{on} is the number of days that the transformer are running, N_{tot} is the total number of days in the corresponding month. For actual maximum demand charge, basic electricity charge is computed as:

$$cost = C_{actual} \times \frac{N_{on}}{N_{tot}} \times 28.5 \tag{2}$$

where C_{actual} is the actual maximum demand. For contract maximum demand charge, basic electricity charge is computed as:

$$cost = \begin{cases} C_{contra} \times \frac{N_{on}}{N_{tot}} \times 28.5, & C_{actual} \leq 1.05 \times C_{contra} \\ [C_{contra} + 2 \times (C_{actual} - 1.05 \times C_{contra})] \times \frac{N_{on}}{N_{tot}} \times 28.5, & others \end{cases} \tag{3}$$

where C_{contra} represents the contract maximum demand. Maximum load prediction helps select the optimal basic electricity charge method and contact maximum demand, and therefore cut down the cost.

Power Rate Electricity Charge. Power rate electricity charge is an additional charge based on the assessment of power factor. For industrial users, the assessment criteria is 0.9, and for other users the criteria is 0.85. If a user's power factor is higher than the criteria, he will be rewarded by a negative power rate electricity charge. Otherwise he will be penalized by a positive one. Therefore, a low power factor also leads to increase of cost.

4 Huffman Tree Based Multi-resolution Temporal Convolution Network

The Section first describes the overall framework of HM-TCN, and then introduces the temporal convolution layer and the Huffman tree based multi-resolution information extraction respectively.

4.1 Overall Framwork

Typical time series prediction methods need to extract information from the time series through feature engineering, and then establish mathematical statistical modeling according to the features obtained. The parameters of these methods need to be determined manually. Besides, the recurrent neural network is used to establish deep learning models. Because the recurrent neural network needs to get inputs cyclically, the training time and the model inference time consumption increased, which can not meet the demand for in the industrial environment.

Observe the above difficulties, we proposes a multi-resolution temporal convolution network based on Huffman tree (HM-TCN). The network can compress and extract temporal information by convolution, and fuse the multi-resolution extracted information by Huffman tree, so as to realize the extraction and aggregation of coarse-grained temporal information. HM-TCN can monitor richer feature information at different resolutions and improve the accuracy of model inference. In addition, because the model uses convolution layers instead of recurrent neural networks, it can train the model and infer the model in parallel, which greatly reduces the time of model training and inference, and can well meet the needs of time series prediction in industrial environments.

The framework of HM-TCN is shown in Fig. 1, which consists of two main parts, temporal convolution layers and a Huffman tree multi-resolution extraction module. The temporal convolution layer is used to extract the temporal feature information of different resolutions, and the correlation information of adjacent moments is obtained by applying the convolution in the time dimension. In order to satisfy the causal property of time series, causal convolution is employed. In addition, with the increase of the number of layers the convolution

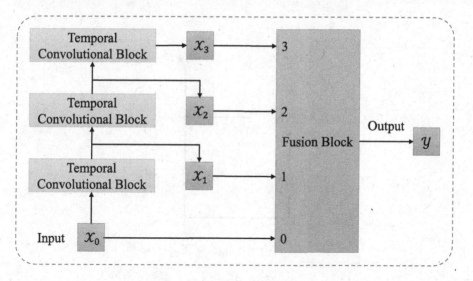

Fig. 1. Framework of HM-TCN

operation increases exponential. We use the dilated convolution to reduce the convolution operation and enable deep layers. Through the dilated convolution structure, we reduce the convolution operation without reducing the convolution field of view, that is, we still extract the same rich information as before. Each convolution layer extracts information based on the convolution operation of the previous layer, so that different convolution layers obtain different resolution of information. The low-level convolution layers obtain more fine-grained information, and the high-level convolution layers obtain more coarse-grained information. The information of different resolution will be integrated to obtain richer feature information at different resolutions, which is called the Huffman tree based multi-resolution extraction. In this way, the output synthesizes the temporal correlation information and the feature information of different resolutions to obtain more accurate results.

We proceed to introduce the temporal convolution layers and Huffman tree multi-resolution information extraction module respectively.

4.2   Temporal Convolution Layers

The temporal convolution layer is mainly used to extract the dependencies and features of adjacent time series, which is mainly composed of causal convolution and dilated convolution, as shown in Fig. 2. Causal convolution considers the causal characteristics of time series, and dilated convolution is used to reduce the convolution operation between each layer, which makes it possible to stack the deep network structure. The two types of convolution are described in detail as follows.

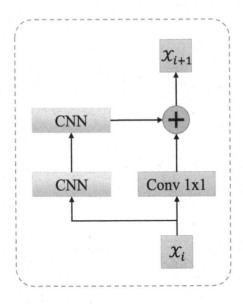

Fig. 2. Convolution Block

Causal Convolution. Causal convolution is a convolution operation performed to satisfy the causality of time series. Causality of time series means that we cannot obtain the information in the future. Therefore, when predicting the value at time t, we cannot use the information at time $t + 1$ in the future. Therefore, when performing conclusion operation on time series, we can only involve the data at time t and before. Besides, in order to keep the output sequence consistent with the input sequence, we use zero padding to fill the series.

The causal convolution convolves the adjacent data, which takes too many computations. For the causal convolution of a long-term sequence, many convolution computations are performed. With the increase of the number of network layers, the convolution operation is doubled, which limits the depth of the network, where only deep network structure can extract thorough information. Therefore, we adopt the operation of dilate convolution to avoid this problem.

Dilate Convolution. Causal convolution performs convolution operations on adjacent temporal data to extract information, which makes it challenging to deal with long sequence data. *Aron* et al. [15] proposed the dilate convolution in 2016, which can obtain an exponentially large field of view. Formally, for a 1-D input $x \in R$ and a filter function $f : 0, \ldots, k - 1 \rightarrow R$, the dilate convolution F on element s is defined as

$$F(s) = \sum_{i=0}^{k-1} f(i) \cdot x_{s-d \cdot i}$$

where d is the dilate factor, k denotes the filter size and $s - d \cdot i$ means accounting for the direction of the past. When $d = 1$, dilate convolution degrade into the common convolution. Using convolution with larger dilate allows neurons at higher levels to capture convolution information with a larger field of view, making deep convolution networks possible.

4.3   Huffman Tree Based Multi-resolution Information Extraction

The temporal convolution layers extract the dependencies and feature information between time series through convolution layer-by-layer, wherein the lower convolution layers extract the data features of adjacent time information, and the upper layers obtain more macroscopic feature information. The previous studies often use the convolution result of the upper layer as the output information directly. However, finer-grained information is lost in this way. For time series prediction, the data near the current moment is more relevant to the data we would like to predict next. And only the macroscopic information of the upper layer is used for prediction, which loses more fine-grained information.

In order to solve the problem, we propose the multi-resolution information extraction based on Huffman tree, which fuses the information of different resolutions in each layer. Through the gradual fusion from high resolution to low resolution, more abundant and accurate information is obtained to predict results.

Huffman tree [12] is a data structure with encoding compression function. During encoding, Huffman tree adopts shorter encoding for higher frequency characters, and longer encoding for lower frequency characters to achieve data compression.

The Huffman tree based multi-resolution feature fusion is shown in the Fig. 3. Firstly, the high-resolution information layer is fused to process the high-resolution information, and then the low-resolution information is fused next. The low-resolution level contains more macroscopic information and has a larger time scale vision. The fusion from high resolution level to low resolution level achieves the effect of knowledge distillation, distilling the feature information we need from the bottom to the top step by step. By integrating the micro information into the macro information, it realizes an all-round multi-resolution information perception.

5   Electricity Cost Optimization

The optimization of charging method is mainly aimed at industrial users, and the optimization scheme can be divided according to the optimization target electricity charge (electricity charge, basic electricity charge, power rate electricity charge). Among them, the optimization of power rate electricity charge mainly depends on offline inspection, and we only need to filter the users with higher power rate and electricity cost. Here we focus on the electricity cost optimization scheme and the basic electricity cost optimization scheme.

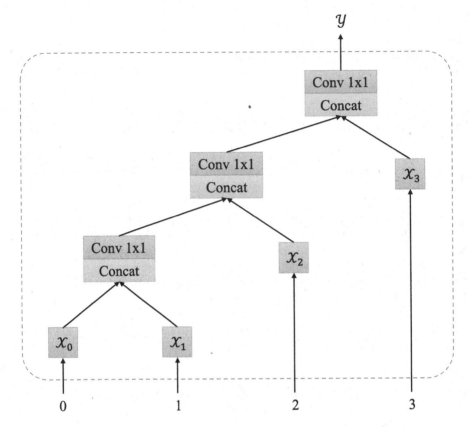

Fig. 3. Huffman Multi-resolution fusion

5.1 Optimization of Electricity Cost

Since the electricity price is charged according to the peak, plain and valley period, the electricity proportion of the peak, plain and valley can be used to analyze whether the electricity consumption period is reasonable. If the user's electricity consumption is concentrated in the peak period, we suggest to transfer the electricity consumption to the plain and valley period appropriately.

For the market users with high deviation electricity, the price of electricity can be reduced by providing them with accurate electricity prediction technology.

5.2 Optimization of Basic Cost

The basic electricity charge optimization scheme are mainly divided into the switching between capacity-based cost, demand-based cost, actual maximum demand, contract maximum demand cost, single system cost and two-part system cost.

Defined three types of capacities $C_{max}, C_{actual}, C_{contra}$, where C_{max} denotes the transformer capacity, C_{actual} denotes the maximum demand capacity and

C_{contra} denotes the contract demand capacity. And three cost schemes S_1, S_2, S_3 are corresponding to three types of capacities respectively. Compare S_1 with S_2, it is more cost-efficient to choose S_1 when $66\% \cdot C_{max} > C_{actual}$, otherwise S_2 is more efficient. Compare S_2 with S_3, it is more cost-efficient to use S_2 when $C_{actual} < C_{contra}$ or $C_{actual} > 1.075 \cdot C_{contra}$, it is more cost-efficient to choose S_2, otherwise S_3 is more efficient.

The finally type of optimization is mainly aim at users whose transformer capacity is close to 315kVA. For a large industrial user, it converts to an ordinary industrial user after reduces the transformer capacity under 315kVA. If the electricity consumption still meets the demand and total electricity costs may decrease due to save the basic electricity cost, we suggest cutting the transformer capacity to the level of ordinary industrial users. Similarly, ordinary industrial users could consider increase the transformer capacity for the opposite reason. Therefore, for the above two types of users, it is necessary to consider the increase and decrease of transformer capacity for optimization.

6 Experiments

6.1 Experimental Setup

The HM-TCN proposed in this paper is implemented under python 3.6 and pytorch 1.11.0. All experiments were done on a Linux (Ubuntu 18.04.1) machine, which was configured with an NVIDIA RTX A6000 graphics card.

Dataset. We conduct experiments on two public datasets:

- *Electricity*[1] contains 2075259 measurements gathered in a house located in Sceaux (7km of Paris, France) between December 2006 and November 2010 (47 months).
- *DBRace*[2] contains 128156 measurement data (44 months) from January 2018 to August 2021 with a data interval of 15 min, and 3610 pieces of data divided by industry with a data interval of one day.

Baseline. For the average load prediction and maximum load prediction problems, we compare the HM-TCN model with the following three methods:

- *ARIMA* [16]: A statistical time series model, considering the moving average component of time series, performs stationary processing and autoregressive prediction on non-stationary time series data.
- *LSTM* [10]: The RNN-based time series model uses input gates, output gates, and forget gates to update the cell state to deal with long-term and short-term dependencies.
- *TCN* [2]: A CNN-based time series model that uses causal dilated convolutions to replace the temporal dependencies of sequences extracted by RNN.

---

[1] https://archive.ics.uci.edu/ml/datasets/individual+household+electric+power+consumption.

[2] https://www.tipdm.org:10010/#/competition/1481159137780998144/question.

6.2   Experimental Results

Model Effect. This summary mainly introduces the experimental results of the model HM-TCN proposed in this paper and the three comparative models ARIMA, LSTM and TCN on the two datasets Electricity and DBrace. The evaluation index used is the root mean square error. Specifically, the experimental results are shown in the Table 1. From the experimental results of the table, it can be found that, on two datasets Electricity and DBrace, the effect of the model HM-TCN proposed in this paper is significantly better than the other three comparison algorithms. Compared to them, the RMSE of HM-TCN on both datasets drops drastically. Specifically, on the dataset Electricity, the RMSE of HM-TCN is only 0.000001, but the RMSE of the second best TCN is 0.000029. In comparison, OurMode reduces the RMSE by 96.5%. On the dataset DBRace, the RMSE of HM-TCN is only 9.171174, but the RMSE of the second best TCN is 85.465812. In comparison, HM-TCN reduces the RMSE by 89.3%. The reason why the method HM-TCN proposed in this paper can achieve the excellent results as shown in the experimental results is that HM-TCN contains a multi-resolution information extraction module based on Huffman tree, this module gradually extracts information of different resolutions from fine-grained to coarse-grained according to the construction process similar to Huffman tree. Through this method, HM-TCN can distill out the important information from the original data, which plays the role of knowledge distillation, so in the end a very good result was achieved.

Table 1. Comparison with three models

Model	Electricity	DBRace
	RMSE	RMSE
ARIMA	1.048876	32937.878526
LSTM	0.002634	465.382265
TCN	0.000029	85.465812
HM-TCN	**0.000001**	**9.171174**

Visualization Interface. We next show the visualization interface of the power load prediction system. The visualization interface consists of four pages: overall electricity prediction, individual electricity prediction, overall electricity cost overview and individual electricity cost optimization.

Overall Electricity Prediction. The overall power prediction page is shown in the Fig. 4. On the left top is the chart of predicted electricity consumption of different industries. On the left bottom is the electricity consumption distribution curve. On the right top is ratio of different industries' electricity consumption.

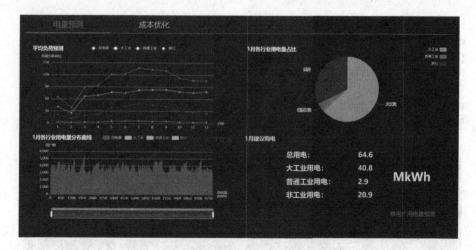

Fig. 4. Overall power prediction

On the right bottom is electricity purchase recommendations for electricity sales companies.

Individual Power Prediction. The individual power prediction interface is shown in the Fig. 5. User details are displayed on the table on the left side of the page. Detail table has the function of searching information by month and username, and sorting by keyword. On the right side, the total, peak, flat, and valley predicted power of each month for the users selected in the table, as well as the power ratio are displayed.

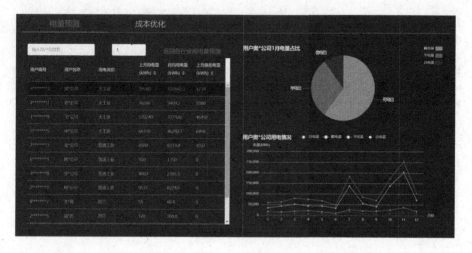

Fig. 5. Individual power prediction

Overall Electricity Cost Overview. An overview of the overall electricity cost is shown in the Fig. 6. The left side of the page shows the monthly average price of electricity and the distribution curve of the average price of electricity in the past few months by industry. By selecting certain interval in the distribution graph, target users with higher average purchase price are filter out in the user table on the right. Single-user electricity consumption details page can be achieved through the bottom in the table.

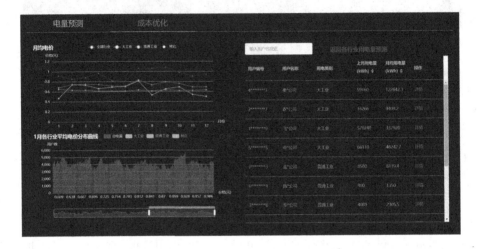

Fig. 6. Overall electricity cost overview

Individual Electricity Cost Optimization. As the Fig. 7 shows, this page displays the detailed information card of the selected user, the proportion of various electricity costs, the predicted electricity situation, the predicted electricity proportion, the maximum load prediction result and the power factor assessment result. You can view the system's cost optimization suggestions for the user through a pop-up window.

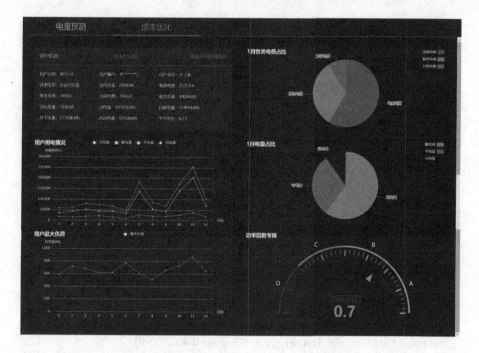

Fig. 7. Individual electricity cost optimization

7 Conclusion

We use deep learning techniques to study the average load prediction problem and the maximum load prediction problem under the background of industrial electricity marketization. Specifically, we propose the HM-TCN model to predict average and maximum loads, and design a set of optimal electricity cost scheme selection strategy that can select the optimal cost scheme based on the load prediction results to save electricity cost, and implement a visualization interface to display electricity information, prediction results, and electricity cost optimization suggestions. We test the proposed model on two public datasets. Experiments show that our proposed model has the lowest RMSE among all four models. Specifically, the RMSE of HM-TCN on both datasets Electricity and DBrace is lower than the second lowest RMSE, decreasing by 96.5% and 89.3%, respectively.

References

1. Al-Hamadi, H., Soliman, S.: Short-term electric load forecasting based on Kalman filtering algorithm with moving window weather and load model. Electr. Power Syst. Res. **68**(1), 47–59 (2004)
2. Bai, S., Kolter, J.Z., Koltun, V.: An empirical evaluation of generic convolutional and recurrent networks for sequence modeling. arXiv preprint. arXiv:1803.01271 (2018)

3. Bendaoud, N.M.M., Farah, N., Ahmed, S.B.: Comparing generative adversarial networks architectures for electricity demand forecasting. Energy Build. **247**, 111152 (2021)

4. Brodersen, K.H., Gallusser, F., Koehler, J., Remy, N., Scott, S.L.: Inferring causal impact using bayesian structural time-series models. Annals Appl. Stat. 247–274 (2015)

5. Chatfield, C.: The holt-winters forecasting procedure. J. Roy. Stat. Soc.: Ser. C (Appl. Stat.) **27**, 264–279 (1978)

6. Cheng, Y.Y., Chan, P.P., Qiu, Z.W.: Random forest based ensemble system for short term load forecasting. In: 2012 International Conference on Machine Learning and Cybernetics, vol. 1, pp. 52–56. IEEE (2012)

7. Chung, J., Gulcehre, C., Cho, K., Bengio, Y.: Empirical evaluation of gated recurrent neural networks on sequence modeling. arXiv preprint. arXiv:1412.3555 (2014)

8. El Desouky, A., El Kateb, M.: Hybrid adaptive techniques for electric-load forecast using ANN and ARIMA. IEE Proc.-Gener. Trans. Distrib. **147**(4), 213–217 (2000)

9. Elman, J.L.: Finding structure in time. Cogn. Sci. **14**, 179–211 (1990)

10. Hochreiter, S., Schmidhuber, J.: Long short-term memory. Neural Comput. **9**(8), 1735–1780 (1997)

11. Huang, Y., Weng, Y., Yu, S., Chen, X.: Diffusion convolutional recurrent neural network with rank influence learning for traffic forecasting. In: 2019 18th IEEE International Conference on Trust, Security and Privacy in Computing and Communications/13th IEEE International Conference on Big Data Science and Engineering (TrustCom/BigDataSE), pp. 678–685. IEEE (2019)

12. Huffman, D.A.: A method for the construction of minimum-redundancy codes. IRE **40**, 1098–1101 (1952)

13. Mariella, L., Tarantino, M.: Spatial temporal conditional auto-regressive model: a new autoregressive matrix. Austrian J. Stat. **39**(3), 223–244 (2010)

14. Mohandes, M.: Support vector machines for short-term electrical load forecasting. Int. J. Energy Res. **26**(4), 335–345 (2002)

15. Oord, A.V.D., et al.: Wavenet: a generative model for raw audio. In: SSW, p. 125 (2016)

16. Pan, B., Demiryurek, U., Shahabi, C.: Utilizing real-world transportation data for accurate traffic prediction. In: 2012 IEEE 12th International Conference on Data Mining, pp. 595–604. IEEE (2012)

17. Rahman, A., Srikumar, V., Smith, A.D.: Predicting electricity consumption for commercial and residential buildings using deep recurrent neural networks. Appl. Energy **212**, 372–385 (2018)

18. Sejnowski, T.J., Rosenberg, C.R.: Parallel networks that learn to pronounce English text. Complex Syst. **1**, 145–168 (1987)

19. Shi, G., Wu, Y., Liu, J., Wan, S., Wang, W., Lu, T.: Incremental few-shot semantic segmentation via embedding adaptive-update and hyper-class representation. In: Proceedings of the 30th ACM International Conference on Multimedia, pp. 5547–5556 (2022)

20. Taylor, J.W., Buizza, R.: Neural network load forecasting with weather ensemble predictions. IEEE Trans. Power Syst. **17**(3), 626–632 (2002)

21. Waibel, A., Hanazawa, T., Hinton, G., Shikano, K., Lang, K.J.: Phoneme recognition using time-delay neural networks. IEEE **37**, 328–339 (1989)

22. Wang, Z., Oates, T.: Imaging time-series to improve classification and imputation. In: 24th International Joint Conference on Artificial Intelligence (2015)

23. Werbos, P.J.: Backpropagation through time: what it does and how to do it. IEEE **78**, 1550–1560 (1990)

24. Wu, Y., Guo, H., Chakraborty, C., Khosravi, M., Berretti, S., Wan, S.: Edge computing driven low-light image dynamic enhancement for object detection. IEEE Trans. Netw. Sci. Eng. (2022)
25. Wu, Y., Zhang, L., Berretti, S., Wan, S.: Medical image encryption by content-aware DNA computing for secure healthcare. IEEE Trans. Ind. Inf. **19**(2), 2089–2098 (2022)
26. Zeger, S.L., Qaqish, B.: Markov regression models for time series: a quasi-likelihood approach. Biometrics **44**, 1019–1031 (1988)
27. Zhang, J., Zheng, Y., Qi, D.: Deep spatio-temporal residual networks for citywide crowd flows prediction. In: 31st AAAI Conference on Artificial intelligence (2017)

Deep Learning-Based Autonomous Cow Detection for Smart Livestock Farming

Yongliang Qiao[1], Yangyang Guo[2(✉)], and Dongjian He[3]

[1] Australian Centre for Field Robotics (ACFR), Faculty of Engineering, The University of Sydney, Sydney, NSW 2006, Australia
[2] School of Internet, Anhui University, Hefei 230039, Anhui, China
guoyangyang113529@ahu.edu.cn
[3] College of Mechanical and Electronic Engineering, Northwest A&F University, Yangling 712100, Shaanxi, China

Abstract. Animal sourced protein is increasing rapidly due to the growing of population and incomes. The big data, robots and smart sensing technologies have brought the autonomous robotic system to the smart farming that enhance productivity and efficiency. Therefore, a YOLOv4-SAM was proposed to achieve high detection precision of cow body parts in long-term complex scenes. The proposed YOLOv4-SAM consists of two components: YOLOv4 is for multi-scale feature extraction, while the Spatial Attention Mechanisms (SAM) highlights the key cow biometric-related features. By doing this, visual biometric feature representation ability is enhanced for improving cow detection performance. To verify the performance of YOLOv4-SAM, a challenging dataset consisting of adult cows and calves with complex environments (e.g., day and night, occlusion, multiple target) was constructed for experimental testing. The precision, recall, mIoU and of the proposed YOLOv4-CBAM were 92.29%, 96.51%, 77.22% and 93.13%, respectively. The data shows that its overall performance was better than that of the comparison algorithm (Faster R-CNN, RetinaNet and YOLOv4). In addition, object detection based on the YOLOv4-SAM model can capture the key biometric-related features for cow visual representation and improve the performance of cow detection. In addition, the detected height difference between head and leg proved the capability in automatic identification of lame cows. The proposed deep learning-based cow detection approach provides a basis for developing an automated system for animal monitoring and management on commercial dairy farms.

Keywords: Autonomous detection · YOLOv4 · Attention mechanism · Deep learning · Precision livestock farming

1 Introduction

Precision farming is key for producing more food for increasing world population, especially the livestock production that provides valuable protein [1]. To meet human demand for animal products, the animal production needs to improve its efficiency and operations for higher productivity [2]. In the past decades, intelligent mechanical equipment and robotics have been one of the main frame-works which focusing on minimizing environmental impacts and simultaneously maximizing agricultural productivity [3, 4]. Automated systems with smart sensors, big data processing based on artificial intelligence, and robots can be used to enhance the efficiency of production such as crop harvesting, fruit picking and animal farming [5, 6].

In smart agriculture, the automatic, efficient and accurate acquisition of animal information has been a key prerequisite. The vision-based animal monitoring, body parts detection, and welfare evaluation can be completed automatically without stress [2]. More recently, deep learning can more effectively realize visual feature-based object detection and behavior recognition with its network depth and feature representation ability [7]. Riekert et al. [11] combined Faster R-CNN and Neural Architecture Search (NAS) to construct a pig position detection and pose recognition system, and verified the feasibility of their methods. Shao et al. [12] designed a cattle detection system based on Convolutional Neural Networks (CNNs), obtained a precision of 0.957. Jiang et al. [13] proposed FLYOLOv3 method to detect body parts (e.g., head, leg and trunk) of individual dairy cow. Although the above approaches demonstrated the feasibility of deep learning-based animal detection, it is still challenging to achieve accurate detecting of key areas of cow in a complex farm environment (e.g., multi-cows, occlusion, different illumination, day and night) [14].

More recently, YOLO network has been a popular method for object detection [18]. In YOLO family, YOLOv4 is better than YOLOv1, YOLOv2 and YOLOv3, and has excellent speed and accuracy [19]. YOLOv4 is an end-to-end real-time deep learning-based method of object detection proposed in 2020, and it has been proved with high detection accuracy and speed on many datasets [19]. In addition, the attention mechanism can improve the extraction ability of target-related features and reduce the interference of non-target area features, thereby improving the effect of target detection models and is widely used [20, 21].

To improve remote animal detection accuracy, we proposed YOLOv4-SAM model to detect the cow. Firstly, YOLOv4 was used to extract multi-scale features from sample images. Then, Spatial Attention Mechanisms (SAM) was used to highlight the animal biometric-related features and enhance the animal detection performance. In this study, images of dairy adult cows and calves during the day and night were acquired for testing. In our detection experiments, whole animal, head, and four legs were detected respectively. And a further qualitative analysis of lame recognition was conducted based on height difference of detected legs and head.

The contributions of this paper include: (1) We integrated the SAM attention mechanism into YOLOv4 and proposed YOLOv4-SAM based approach for cow boy part detection. The proposed YOLOv4-SAM improves the biometric feature representation ability and enhances cow body part detection performance; (2) A long-term challenging dataset consisting of adult cow and calf with complex environments (e.g., day and night, occlusion, multiple target) was constructed for detection verification. The results indicate that the YOLOv4-SAM method was superior to the comparison method, with high detection accuracy and fast processing speed, which satisfy the real-time requirement of smart farm applications; and (3) Th height difference of YOLOv4-SAM based leg and head detection could further used to detect lameness. Overall, the proposed

YOLOv4-SAM provides a real-time and high accuracy cow body part detection approach in complex scenes, which facilitates long-term autonomous animal detection and health/welfare evaluation.

2 Material and Methods

2.1 Data Acquisition

In this experiment, the data sets for calf and adult cattle were sourced from Yangling Keyuan Cloning Co., Ltd. China. For calf data, a Holstein calf was monitored in a rectangular enclosure (4 m × 2 m × 1.5 m). The height of the installed camera was 0.75 m and the distance from the fence is 2.5 m. For adult cow data, the camera was placed on a support beam of the 1.8-m-high feed shed, and it was 35 m away from the aisle. Camera setup for calf and cow video recording is presented in Fig. 1.

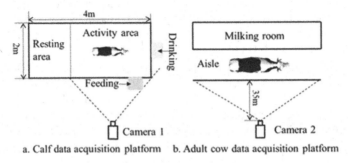

a. Calf data acquisition platform   b. Adult cow data acquisition platform

Fig. 1. Video acquisition setup diagram of cattle.

The dataset acquired consider is really challenging due to the factors: multi-cow appeared and exist occlusion; Lighting is changing from dawn to dusk; the complex background including crush, soil ground, building background and so on. The frame rate, bit rate and resolution of the captured cattle videos were 25 fps, 2000 kbps, 704 pixels × 576 pixels respectively. A total of 1040 images were obtained, including 540 images of adult cows and 500 images of calf. Among these images, 600 and 440 images were randomly selected as the training and testing datasets. In our experiments, the whole cow, the head, and legs (left front leg, left hind leg, right front leg, and right hind leg) were labeled manually with bounding boxes, respectively, for key body areas detection testing.

2.2 YOLOV4-SAM Model for Detecting Dairy Cow

In order to improve the detection performance for key cow parts, we integrated SAM attention module into the YOLOv4 object detector, which had higher accuracy in cow detection and monitoring.

The proposed YOLOv4-SAM approach extracted multi-scale features using YOLO4 object detector and learning feature importance through SAM, enhancing the representation of animal biometric related features and improving the animal detection performance. The proposed YOLOv4-SAM-based detection method includes four parts, backbone, neck, SAM, and head as shown in Fig. 2.

- CSPDarknet53 as backbone network of YOLOv4 algorithm was be used to extract target features [19].
- The Neck was composed of the PANet and SPPNet. PANet proposed a bottom-up information propagation path enhancement method, which realized bottom-up feature extraction through convolution and up-sampling, and realized top-down feature extraction through down-sampling, thereby better fusing the extracted features [22]. The SPPNet module can concatenate feature maps from different core sizes together as an output, effectively increasing the backbone's acceptance domain and separating significant context features [23].
- SAM was used to enhance the weight of the target candidate region after the convolution block [24].
- And then the YOLOv3 head was used to realize object detection [25].

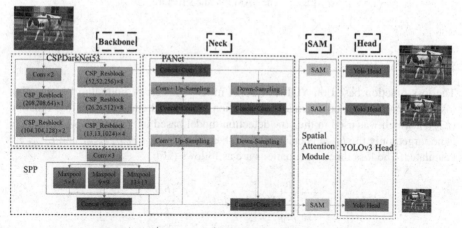

Fig. 2. YOLOv4-SAM based cow detection framework.

2.2.1 SAM Module for Feature Optimization

The SAM mainly encoded spatial pixel correlations in the feature map into local features to enhance their representation ability. The SAM module structure is illustrated in Fig. 3. In the feature map $F_t \in R^{C \times H \times W}$, C, H and W represent the channel number, height, and width, respectively. Pooling operator includes the max pooling and the average pooling. The equation is as follows:

$$P_a = AvePool(F_t) \tag{1}$$

$$P_m = MaxPool(F_t) \tag{2}$$

where *AvePool(·)* and *MaxPool(·)* indicate average pool and max pool, respectively.

Then P_a and P_m form a new feature P_{am} by *Concat()* function. After that, the spatial attention map *As* is obtained by the *Conv ()* function and the *Sigmoid (·)* function:

$$A_s = Sigmoid(Conv(Concat(P_a, P_m))) \tag{3}$$

Finally, the feature map Fs $\in R^{C \times H \times W}$ can be calculated as:

$$F_s = A_s \otimes F_t \tag{4}$$

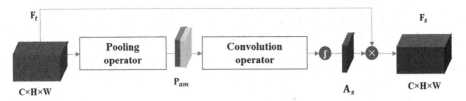

Fig. 3. The SAM module structure.

2.3 Loss Function

The loss function based on YOLOv4-SAM model consists of bounding box location loss function (L_{CIoU}), confidence loss function (L_{conf}) and classification loss function (L_{class}), which was used in the cow detection model based on YOLOv4-SAM. If there is no target, only the L_{conf} is calculated. If there is a target, L_{CIoU}, L_{conf} and L_{class} are calculated. The loss function equations are as follows [26]:

$$L_{CIoU} = \sum_{i=0}^{S^2} \sum_{j=0}^{B} I_{i,j}^{obj} \left[1 - IoU + \frac{\rho^2(b, b^{gt})}{c^2} + \alpha v \right] \tag{5}$$

$$v = \frac{4}{\pi^2} \left(arctan \frac{w^{gt}}{h^{gt}} - arctan \frac{w}{h} \right)^2 \tag{6}$$

$$\alpha = \frac{v}{((1 - IoU) - v)} \tag{7}$$

$$L_{conf} = -\sum_{i=0}^{S^2} \sum_{j=0}^{B} I_{i,j}^{obj} \left[\hat{C}_i \log(C_i) + \left(1 - \hat{C}_i\right) \log(1 - C_i) \right]$$

$$- \lambda_{noobj} \sum_{i=0}^{S^2} \sum_{j=0}^{B} I_{i,j}^{noobj} \left[\hat{C}_i \log(C_i) + \left(1 - \hat{C}_i\right) \log(1 - C_i) \right] \tag{8}$$

$$L_{class} = -\sum_{i=0}^{S^2} I_{i,j}^{obj} \sum_{c \in classes} \left[\hat{P}_i(c) \log(P_i(c)) + \left(1 - \hat{P}_i(c)\right) \log(1 - P_i(c)) \right] \quad (9)$$

$$Loss = L_{CIoU} + L_{conf} + L_{class} \quad (10)$$

where, S is the number of grids, B is the number of prior boxes in each grid. λ_{noobj} represents weight. $I_{i,j}^{obj}$ and $I_{i,j}^{noobj}$ are used to determine whether the $j-th$ priori box of the $i-th$ grid contains the object. If yes, $I_{i,j}^{obj}$ is 1 and $I_{i,j}^{noobj}$ is 0. Otherwise $I_{i,j}^{obj}$ is 0 and $I_{i,j}^{noobj}$ is 1. IoU refers to the ratio of intersection and union of the prediction and actual bounding boxes. ρ () is the Euclidean distance, and c is the diagonal distance between the predicted box and the closure area of actual box. b, w and h are the center coordinates, width and height of the prediction box, respectively. b^{gt}, w^{gt} and h^{gt} are the center coordinates, width and height of the actual box, respectively. α is weight factor, v is similarity ratio of length to width. C_i means the confidence of prediction and label box. $P_i(c)$ is the classification probability of different categories; c is the number of detection categories.

3 Experimental Setup

3.1 The Used Dataset

The sample descriptions of training and testing sets are shown in Table 1.

Table 1. Datasets description in the experiments

Dataset	Number of datasets
Train	Adult cow: 300 (270 days; 30 nights)
	Calf: 300 (270 days; 30 nights)
Test	Adult cow: 240 (220 days; 20 nights)
	Calf: 200 (180 days; 20 nights)

3.2 Network Training Platform

In our work, all data analysis works were carried out on a computer equipped with a GeForce GTX 1080 Ti GPU and I9-7920X CPU@2.9 GHz, using Keras framework. To verify the effectiveness of the dairy cow key parts detection model, Faster R-CNN [27], RetinaNet [28] and YOLOv4 [19] were used for comparison. In addition, the input size of all networks was $416 \times 416 \times 3$, epoch was 1000, batch size was 16, and learning rate was 0.0013. All the initial weights of the network were random. The other parameters were the default settings.

4 Results

4.1 Detection Performance Comparison

The proposed YOLOv4-SAM based cow detection was compared to other CNN based approaches in Table 2. It can be seen that the precision and recall of the YOLOv4-SAM approach is up to 92.29% and 96.51%, respectively, which is higher than that of Faster R-CNN (53.85%,63.25%), RetinaNet (74.95%,75.69%) and YOLOv4 (91.86%,96.51%). For the mIoU, YOLOv4-SAM method is 77.22%, higher than that of Faster R-CNN (51.36%), RetinaNet (76.36%) and YOLOv4 (75.18%). In addition, it can be noticed that the mAP@0.5 of YOLOv4-SAM is 93.13%, which is higher than that of YOLOv4 (93.08%). These results demonstrated that the proposed YOLOv4-SAM approach can pay more attention to the visual features related to the animal body and enhance the animal detection ability.

In addition, the detection speed of our proposed YOLOv4-SAM (40 f/s) is fast than that of Faster-RCNN and RetinaNet networks, and lower than the original YOLOv4 network. But the average recognition speed could reach 40 FPS, which would highly possible to satisfy the real-time requirement (>20 FPS) of robotic based autonomous cow detection. Overall, the proposed YOLOv4-SAM highlights the related animal detection related features and enhances the detection performance, which provides a favorable solution for the remote animal detection in smart livestock farming.

Table 2. Comparison of different GDM methods.

Method	Precision (%)	Recall (%)	mIoU (%)	mAP@0.5 (%)	FPS (f/s)
Faster R-CNN	53.85	63.25	51.36	59.72	27
RetinaNet	74.95	75.69	76.36	74.38	29
YOLOv4	91.86	96.51	75.18	93.08	55
YOLOv4-SAM	**92.29**	**96.51**	**77.22**	**93.13**	**40**

Cow and calf images from the different scenes (e.g., day and night, occlusion, multiple target) were selected to evaluate the performance of the model. Figure 4 shows the detection results of our proposed YOLOv4-SAM approach. It shows that the YOLOv4-SAM approach could recognize the images of cow and calf at night, and accurately detect the head and legs, which is beneficial to the long-term cattle detection in smart animal husbandry. All this indicating that robustness of the proposed YOLOv4-SAM approach.

Fig. 4. Examples of YOLOv4-SAM based cow and calf detection. In the detection result, the orange, green, purple, light green, yellow, red and blue boxes are adult cow area, claf area, head, outer front leg, inner front leg, outer hind leg and inner hind leg, respectively. (Color figure online)

4.2 Detection Results of Key Body Parts of Cattle

To further analysis the detection performance of different body parts, in Table 3, the numbers of detected body parts (including tp and fp), precision, recall, IoU and mAP@0.5 are presented.

The correctly detected number (tp) and false detected number (fp) of YOLOv4-SAM are 2759 and 216 respectively, and the tp is 2757 and fp is 260 in YOLOv4. The overall precision, recall, and map@0.5 of YOLOv4-SAM are 92.29%, 96.51% and 93.13% respectively, which higher than that of YOLOv4. In animal detection, the proposed YOLOv4-SAM achieved 96.95% and 99.00% precision for the whole body of adult cow and calf, respectively. As the animal body truck is large than other body parts (e.g., head, leg), both YOLOv4 and YOLOv4-SAM achieved higher detection accuracies for animal itself (e.g., adult cow and calf), and the detection recall of the YOLOv4-SAM approach is more noticeable. From these data, it is clear that the SAM module in YOLOv4 improves the animal visual biometric feature (e.g., shape, contour, and coat color) representation ability, enhancing the cow detection in different environments.

Table 3. Comparison of different body part detection performance.

Method	Species	GT	Detect (tp)	Detect (fp)	Precision (%)	Recall (%)	IoU (%)	mAP@0.5 (%)
YOLOv4	Adult cow	286	286	9	96.95	100.00	85.75	99.94
	Calf	199	199	6	97.07	100.00	89.02	99.68
	Head	481	471	11	97.73	98.54	80.19	90.91
	Outer front leg	482	460	47	90.73	95.44	72.04	90.63

(continued)

Table 3. (*continued*)

Method	Species	GT	Detect (tp)	Detect (fp)	Precision (%)	Recall (%)	IoU (%)	mAP@0.5 (%)
	Inter front leg	472	454	36	92.65	96.19	73.38	90.57
	Outer hind leg	479	456	67	87.19	95.20	71.06	90.51
	Inner hind leg	465	431	84	83.69	92.69	66.44	89.36
	All	**2855**	**2757**	**260**	**91.86**	**96.51**	**75.18**	**93.08**
YOLOv4-SAM	Adult cow	286	286	9	96.95	100.00	90.52	99.87
	Calf	199	199	12	99.00	100.00	92.62	99.82
	Head	481	476	8	98.35	98.96	82.15	90.91
	Outer front leg	482	458	33	93.28	95.02	75.00	90.51
	Inter front leg	472	460	34	93.12	97.46	75.16	90.67
	Outer hind leg	479	454	61	88.16	94.78	73.85	90.52
	Inner hind leg	465	435	69	86.31	93.55	67.75	89.58
	All	**2864**	**2759**	**216**	**92.29**	**96.51**	**77.22**	**93.13**

In addition, the detection results of Table 3 show that the precision of the YOLOv4-SAM for head (98.35%), outer front leg (93.28%), inter front leg (93.12%), outer hind leg (88.16%) and inner hind leg (86.31%) is higher than that of YOLOv4 (97.73% for head, 90.73% for outer front leg, 92.65% for inter front leg, 87.19% for outer hind leg and 863.69% for inner hind leg). It also can be noticed that head detection precision is higher than that of leg, the main reason is that cow head account for a large area and not occluded by other body parts, thus the extracted visual features from head has few disturbing factors. All these increased values show that the application of SAM was feasible in the cow body parts' detection.

4.3   Application of Lame Detection in Dairy Cows

Cow lameness can reflect the health problems of cattle, and the detection of abnormal behavior is important for farms [29]. When a cow is lame, its head position will be lower than that of a normal cow.

To further explored the relative height of the head and leg of the lame cow and the normal cow. Firstly, 30 images of normal cows and lame cows were selected, respectively. Then, yolov4-SAM model was used to detect the head and legs of cows. The y in the central coordinate *(x, y)* of the bounding box of the head was used to represent the height information of the head. Similarly, y_i in the central coordinate *(x_i, y_i)* of the bounding box of the leg was obtained, *i* represented the number of legs and the average value y_{avr} of y_i was taken as the height information of the legs. The difference between y and y_{avr} was denoted as relative height of the head position and the leg position (Fig. 5).

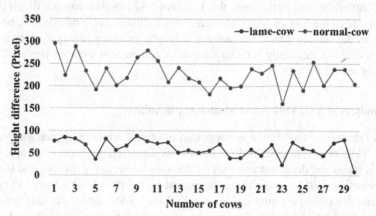

Fig. 5. The relative height difference between the head and leg of normal cows and lame cows.

It can be seen from Fig. 5 that the height difference between the head and leg of lame cows (average is 60.6 pixel) is lower than that of normal cows (average is 165.9 pixel). That means using the position of cow head and leg, namely, the height difference, cow lameness behavior could be detected.

5 Discussion

5.1 Impact of Object Size and Number on Detection Performance

As the YOLOv4 using bounding box to detect object, detecting object with different sizes could have varying performance. The whole-cow body account for a large proportion result in high detection accuracy, while the leg area or head accounted for the small proportion obtained lower detection precision. As illustrated in Table 3, the whole cow detection accuracy was higher than that of head and leg. Although there was a detection difference between whole-cow body, head, and leg, the overall detection performance was over 90% due to the multi-scale feature extraction ability in YOLOv4-SAM.

In Fig. 4, when there is a part of the cow in the scene; or when multiple cows occlude each other, the proposed YOLOv4-SAM could still well detect. This is because that multi-scale feature extraction ability of YOLOv4 and the key feature highlighting mechanism of SAM attention module, visual biometric related features could be well extracted for cow detection.

5.2   Impact of Different Scenes on Detection Performance

To verify the robustness of the YOLOv4-SAM method, the experimental samples contain data from different scenes, as shown in Fig. 4. It shows that the whole body, head and legs of adult cows and calves in different scenes (e.g., day and night, occlusion, multiple target) are well detected. In addition, the adult cow samples include railings, which will occlude part of the legs. In this case, the YOLOv4-SAM method can still detect the legs well, but due to the loss of some information, it will also cause misrecognition. From Table 2, in the sample data set containing multiple scenarios, the performance of the YOLOv4-SAM in this study is better than other comparison methods (Faster R-CNN, RetinaNet and YOLOv4).

5.3   Analysis of YOLOv4-SAM Model Applicability

YOLOv4 model is a real-time object detection algorithm, which integrated the characteristics of YOLOv1, YOLOv2, YOLOv3 and other advanced methods, and improved the detection speed and detection accuracy of the object detection network. And the attention mechanism SAM can highlight the animal biometric-related features to enhance the animal detection performance. So, the YOLOv4-SAM model not only retains the performance of YOLOv4, but also improves the recognition ability of the network. In addition, this study further explored the identification of lame cows. It can be seen from Fig. 5 that the head-to-leg height distance of lame cows is lower than that of normal cows, which can be used as a basis for judging the lame cow. This model can also be applied to assist UAV or UGV for monitoring group animals or animal body parts (e.g., heads, legs, and back, etc.). The movement information of an individual animal's head and legs can be further detected for analyzing animal behavior.

6   Conclusions

In this study, a YOLOv4-SAM based approach was proposed to detect cow and its different body parts remotely. The proposed approach integrating attention mechanism SAM into YOLOv4, enhancing animal visual biometric feature representation ability, which provides a new way for long-term and real-time animal detection for further autonomous based smart livestock farming. A challenging cow dataset consisting of adult and calf with complex environments (e.g., day and night, occlusion, multiple target) was constructed for experimental testing. The proposed YOLOv4-SAM based approach outperformed Faster R-CNN, RetinaNet, and YOLOv4. Meanwhile, the detection performance of different cow body was investigated. Experimental results demonstrated that the proposed YOLOv4-SAM approach could capture key biometric-related features for animal visual representation, improving the performance of cow detection. In addition, the detected height difference of head and leg could be further used to identify lame cow from the health group. Overall, the proposed deep learning-based cow detection approach is favorable for autonomous cow management in smart livestock farming.

References

1. Qiao, Y., et al.: Intelligent perception for cattle monitoring: a review for cattle identification, body condition score evaluation, and weight estimation. Comput. Electron. Agric. **185**, 106143 (2021)
2. Fournel, S., Rousseau, A.N., Laberge, B.: Rethinking environment control strategy of confined animal housing systems through precision livestock farming. Biosys. Eng. **155**, 96–123 (2017)
3. Mishra, S., Syed, D.F., Ploughe, M., Zhang, W.: Autonomous vision-guided object collection from water surfaces with a customized multirotor. IEEE/ASME Transactions on Mechatronics **26**, 1914–1922 (2021)
4. Lim, J., Pyo, S., Kim, N., Lee, J., Lee, J.: Obstacle magnification for 2-D collision and occlusion avoidance of autonomous multirotor aerial vehicles. IEEE/ASME Trans. Mechatron. **25**(5), 2428–2436 (2020)
5. Zhang, Z., Kayacan, E., Thompson, B., Chowdhary, G.: High precision control and deep learning-based corn stand counting algorithms for agricultural robot. Auton. Robot. **44**(7), 1289–1302 (2020). https://doi.org/10.1007/s10514-020-09915-y
6. Ding, H., Gao, R.X., Isaksson, A.J., Landers, R.G., Parisini, T., Yuan, Y.: State of AI-based monitoring in smart manufacturing and introduction to focused section. IEEE/ASME Trans. Mechatron. **25**(5), 2143–2154 (2020)
7. Qiao, Y., Truman, M., Sukkarieh, S.: Cattle segmentation and contour extraction based on mask R-CNN for precision livestock farming. Comput. Electron. Agric. **165**, 104958 (2019)
8. Beggs, D., Jongman, E., Hemsworth, P., Fisher, A.: Lame cows on australian dairy farms: a comparison of farmer-identified lameness and formal lameness scoring, and the position of lame cows within the milking order. J. Dairy Sci. **102**(2), 1522–1529 (2019)
9. Okinda, C., et al.: A review on computer vision systems in monitoring of poultry: a welfare perspective. Artif. Intell. Agric. **4**, 184–208 (2020). https://doi.org/10.1016/j.aiia.2020.09.002
10. Li, G., et al.: Practices and applications of convolutional neural network- based computer vision systems in animal farming: a review. Sensors **21**(4), 1492 (2021)
11. Riekert, M., Klein, A., Adrion, F., Hoffmann, C., Gallmann, E.: Automatically detecting pig position and posture by 2D camera imaging and deep learning. Comput. Electron. Agric. **174**, 105391 (2020)
12. Shao, W., Kawakami, R., Yoshihashi, R., You, S., Kawase, H., Naemura, T.: Cattle detection and counting in UAV images based on convolutional neural networks. Int. J. Remote Sens. **41**(1), 31–52 (2020)
13. Jiang, B., Wu, Q., Yin, X., Wu, D., Song, H., He, D.: FLYOLOv3 deep learning for key parts of dairy cow body detection. Comput. Electron. Agric. **166**, 104982 (2019)
14. He, D., Liu, D., Zhao, K.: Review of perceiving animal information and behavior in precision livestock farming. Trans. Chin. Soc. Agric. Mach **47**(5), 231–244 (2016)
15. Mahmud, M.S., Zahid, A., Das, A.K., Muzammil, M., Khan, M.U.: A systematic literature review on deep learning applications for precision cattle farming. Comput. Electron. Agric. **187**, 106313 (2021)
16. Meena, S.D., Agilandeeswari, L.: Smart animal detection and counting framework for monitoring livestock in an autonomous unmanned ground vehicle using restricted supervised learning and image fusion. Neural Process. Lett. **53**(2), 1253–1285 (2021)
17. Barbedo, J.G.A., Koenigkan, L.V., Santos, T.T., Santos, P.M.: A study on the detection of cattle in UAV images using deep learning. Sensors **19**(24), 5436 (2019)
18. Shafiee, M.J., Chywl, B., Li, F., Wong, A.: Fast YOLO: a fast you only look once system for real-time embedded object detection in video. arXiv preprint arXiv:1709.05943 (2017)
19. Bochkovskiy, A., Wang, C.Y., Liao, H.Y.M: YOLOv4: optimal speed and accuracy of object detection. arXiv preprint arXiv:2004.10934 (2020)

20. Fukui, H., Hirakawa, T., Yamashita, T., Fujiyoshi, H.: Attention branch network: learning of attention mechanism for visual explanation. In: Proceedings of the IEEE/CVF Conference on Computer Vision and Pattern Recognition, pp. 10 705–10 714 (2019)

21. Li, X., Jia, X., Wang, Y., Yang, S., Zhao, H., Lee, J.: Industrial remaining useful life prediction by partial observation using deep learning with supervised attention. IEEE/ASME Trans. Mechatron. **25**(5), 2241–2251 (2020)

22. Liu, S., Qi, L., Qin, H., Shi, J., Jia, J.: Path aggregation network for instance segmentation. In: Proceedings of the IEEE Conference on Computer Vision and Pattern Recognition, pp. 8759–8768 (2018)

23. He, K., Zhang, X., Ren, S., Sun, J.: Spatial pyramid pooling in deep convolutional networks for visual recognition. IEEE Trans. Pattern Anal. Mach. Intell. **37**(9), 1904–1916 (2015)

24. Hu, J., Zhi, X., Shi, T., Zhang, W., Cui, Y., Zhao, S.: Pag-YOLO: a portable attention-guided yolo network for small ship detection. Remote Sens. **13**(16), 3059 (2021)

25. Redmon, J., Farhadi, A.: YOLOv3: an incremental improvement. arXiv preprint arXiv:1804. 02767 (2018)

26. Wang, C., Luo, Q., Chen, X., Yi, B., Wang, H.: Citrus recognition based on YOLOv4 neural network. In: Journal of Physics: Conference Series, vol. 1820, no. 1, p. 012163. IOP Publishing (2021)

27. Ren, S., He, K., Girshick, R., Sun, J.: Faster R-CNN: towards real-time object detection with region proposal networks. In: Advances in Neural Information Processing Systems, vol. 28, pp. 91–99 (2015)

28. Lin, T.Y., Goyal, P., Girshick, R., He, K., Dollár, P.: Focal loss for dense object detection. In: Proceedings of the IEEE International Conference on Computer Vision, pp. 2980–2988 (2017)

29. Kaixuan, Z., Dongjian, H.: Target detection method for moving cows based on background subtraction. Int. J. Agric. Biol. Eng. **8**(1), 42–49 (2015)

Author Index

Printed in the United States
by Baker & Taylor Publisher Services